Welding Fundamentals

Welding Fundamentals

Carson Evans

CLANRYE
INTERNATIONAL
www.clanryeinternational.com

Clanrye International,
750 Third Avenue, 9th Floor,
New York, NY 10017, USA

ISBN: 978-1-64726-140-5

Cataloging-in-Publication Data

Welding fundamentals / Carson Evans.
    p. cm.
Includes bibliographical references and index.
ISBN: 978-1-64726-140-5
1. Welding. 2. Electric welding. 3. Metal-work. I. Evans, Carson.
TS227 .W45 2022
671.52--dc23

For information on all Clanrye International publications
visit our website at www.clanryeinternational.com

CLANRYE
INTERNATIONAL

Table of Contents

Preface

It is with great pleasure that I present this book. It has been carefully written after numerous discussions with my peers and other practitioners of the field. I would like to take this opportunity to thank my family and friends who have been extremely supporting at every step in my life.

The fabrication process that is used to join materials such as metals and thermoplastics is referred to as welding. In this sculptural process, heat is used to melt the parts and left to cool that ultimately causes fusion. Various energy sources are used for welding, the most common of them are gas flame, an electric arc, an electron beam, friction and ultrasound. Shielding metal arc welding, flux-cored arc welding, submerged arc welding, electroslag welding, gas metal arc welding and gas tungsten arc welding are the most commonly used welding methods. Significant precautions should be made to avoid injuries during welding such as electric shock, inhalation of poisonous gases and fumes, burns, and intense ultraviolet radiation. The topics included in this book on welding are of utmost significance and bound to provide incredible insights to readers. Different approaches, evaluations and methodologies on various techniques of welding have been included herein. Coherent flow of topics, student-friendly language and extensive use of examples make this book an invaluable source of knowledge.

The chapters below are organized to facilitate a comprehensive understanding of the subject:

Chapter – Introduction

The fabrication process that joins metals or thermoplastics by using heat for melting and cooling for fusion of joints is referred to as welding. Some common types of welding are gas welding, arc welding, resistance welding, radiant energy welding, etc. This is an introductory chapter which will introduce briefly all these types of welding as well as welding hazards.

Chapter – Gas Welding

The process of melting and joining metals by using flame caused by the reaction of fuel gas and oxygen is called gas welding. It can be categorized into oxy fuel welding and oxy hydrogen welding. The topics elaborated in this chapter will help in gaining a better perspective about these types of gas welding as well as welding flames.

Chapter – Arc Welding

Arc welding refers to the welding process that uses electricity for creating heat to melt and join metals. Shielded metal arc welding, flux cored arc welding, gas tungsten arc welding, plasma arc welding, submerged arc welding, etc. are a few of its types. This chapter closely examines these different types of arc welding to provide an extensive understanding of the subject.

Chapter – Solid State Welding

Solid state welding is referred to a welding process which joins the pieces of material by using high pressure and a temperature below the melting point of the parent material. It includes explosive welding, forge welding, friction welding, diffusion welding, etc. This chapter discusses in detail these concepts of solid state welding.

Chapter – Resistance Welding

Resistance welding uses the electrical resistance of materials which results in heat to form the weld. There are various types of resistance welding such as spot welding, flash welding, seam welding, projection welding, etc. This chapter has been carefully written to provide an easy understanding of resistance welding and its types.

Chapter – Welding Defects

The weld surface irregularities and imperfections in welded parts are termed as welding defects. These defects can be classified as weld cracking, weld spatter, solid inclusions, porosity etc. The topics elaborated in this chapter will help in gaining a better perspective about these welding defects.

Carson Evans

1

Introduction

The fabrication process that joins metals or thermoplastics by using heat for melting and cooling for fusion of joints is referred to as welding. Some common types of welding are gas welding, arc welding, resistance welding, radiant energy welding, etc. This is an introductory chapter which will introduce briefly all these types of welding as well as welding hazards.

Welding

"Welding is the process of joining together two pieces of metal so that bonding takes place at their original boundary surfaces". When two parts to be joined are melted together, heat or pressure or both is applied and with or without added metal for formation of metallic bond.

Need for Welding

With ever increasing demand for both high production rates and high precision, fully mechanized or automated welding processes have taken a prominent place in the welding field. The rate at which automation is being introduced into welding process is astonishing and it may be expected that by the end of this century more automated machines than men in welding fabrication units will be found. In addition, computers play critical role in running the automated welding processes and the commands given by the computer will be taken from the programs, which in turn, need algorithms of the welding variables in the form of mathematical equations. To make effective use of the automated systems it is essential that a high degree of confidence be achieved in predicting the weld parameters to attain the desired mechanical strength in welded joints.

To develop mathematical models to accurately predict the weld strength to be fed to the automated welding systems has become more essential.

Classification of Welding Processes

There are many types of welding techniques used to join metals. The welding processes differ in the manner in which temperature and pressure are combined and achieved. The welding process is divided into two major categories: Plastic Welding or Pressure Welding and Fusion Welding or Non-Pressure Welding.

- Plastic Welding or Pressure Welding: When the metal piece acquires plastic state on heating, external pressure is applied. In this process, externally applied forces play an important role in the bonding operation. "A group of welding processes which produces coalescence at temperatures essentially below the melting point of the base materials being joined without the addition of a filler metal" is Pressure Welding Process. Without melting the base metal, due to temperature, time and pressure coalescence is produced. Some of the very oldest processes are included in solid state welding process. The advantage of this process is the base metal does not melt and hence the original properties are retained with the metals being joined.

- Fusion Welding or Non-Pressure Welding: The material at the joint is heated to a molten state and allowed to solidify. In this process the joining operation involves melting and solidification and any external forces applied to the system do not play an active role in producing coalescence. Usually fusion welding uses a filler material to ensure that the joint is filled. All fusion welding processes have three requirements: Heat, Shielding and Filler material.

Welding of Different Materials

The most available metal in the earth's crust is aluminium whereas, steel is the most used metal. In majority of cases aluminium alloys are replacing steels in industrial applications. Aluminium alloys have low density i.e. nearly one third when compared to steels. Some of these materials are allowing for a significant reduction of weight when compared with structural steels. Aluminium alloys are important for the fabrication of components and structures which require high strength, low weight or electric current carrying capabilities to meet their service requirements. The aluminium alloys can resist the oxidation process, corrosion by water and salt, which steel cannot. The most desirable properties of aluminium and its alloys are the light weight, appearance, ability for fabrication, strength and corrosion resistance and hence it is used for wide variety of applications. When used in aerospace, rail and road vehicles these attributes enable energy efficient operation. In the aerospace applications, materials with high strength-to-weight ratio are required such as aluminium alloys. The production of components of aluminium alloys is not very complex; but joining of these materials can sometimes cause serious problems. Among all aluminium alloys, 6XXX plays a major role in the aerospace industry. They are widely used in the aerospace applications because it has good formability, weldability, machinability, corrosion resistance and good strength when compared to other series of aluminium alloys. Hence these alloys are chosen in this work for FSW process.

Comparison of other Processes for Welding Aluminium Alloys

Tungsten inert gas welding (TIG) and Laser beam welding are used for welding aluminium alloys in aerospace welding. TIG has replaced other arc welding processes for joining aluminium alloys. As the fusion welding of aluminium alloys poses problems like porosity, distortion due to high thermal conductivity and solidification shrinkage, it is not preferred. Welding long butt or lap joints of aluminium alloys using conventional welding techniques is difficult as it cannot be made without distortion. Higher welder skills and special procedures are required as all fusion techniques cause loss of alloying elements through evaporation. To avoid the loss through evaporation, special filler material having 5% of silicon is to be used. If AA 6XXX welds are subjected to post weld solutionizing and aging, the mechanical properties can be further improved. As there is increase in welding of aluminium metals in aerospace and other light weight alloys, it leads to development of Friction Stir Welding. As there is no melting during welding, and also the joints are made in the solid state itself, defects are found to be minimum.

Welding Joints

Welding joints is an edge or point where two or more metal pieces or plastic pieces are joined together. The two or more workpieces (either metal or plastic) are joined with the help of a suitable welding process to form a strong joint. According to the American Welding Society, there are basically five types of welding joints and these are Butt,Corner, Lap, Tee and edge joint.

The 5 basic welding joints are:

- Butt joint
- Corner joint
- Lap joint
- Tee joint and
- Edge joint

Butt Joint

The joint which is formed by placing the ends of two parts together is called butt joint. In butt joint the two parts are lie on the same plane or side by side. It is the most simplest type of joint used to join metal or plastic parts together.

The different weld types in butt welding are:

- Square Butt weld
- Bevel groove weld
- V-groove weld
- J-groove weld
- U-groove weld
- Flare-V-groove weld
- Flare-bevel-groove butt weld

Corner Joint

The joint formed by placing the corner of two parts at right angle is called corner joint. Two parts which is going to be weld with corner joint forms the shape of L.

The different weld types in corner joint are as follows:

- Fillet weld
- Spot weld

- Square-groove weld or butt weld
- V-groove weld
- Bevel-groove weld
- U-groove weld
- J-groove weld
- Flare-V-groove weld
- Edge weld
- Corner-flange weld

T-Joint

The joint which is made by intersecting two parts at right angle (i.e at 90 degree) and one part lies at the centre of the other. It is called as T joint as the two part welded look like english letter 'T'.

The types of welds in T joint are as follows:

- Fillet weld
- Plug weld
- Slot weld
- Bevel-groove weld
- J-groove weld
- Flare-bevel groove
- Melt-through weld

Lap Joint

The lap joint is formed when the two parts are placed one over another and than welded. It may one sided or double sided. This types of welding joints are mostly used to join two pieces with different thickness.

The Various weld types in lap joint are:

- Fillet weld
- Bevel-groove weld
- J-groove weld
- Plug weld
- Slot weld

- Spot weld
- Flare-bevel-groove weld

Edge Joint

The joint formed by welding the edges of two parts together are called edge joint. This joint is used where the edges of two sheets are adjacent and are approximately parallel planes at the point of welding. In this joint the weld does not penetrates completely the thickness of joint, so it can not be used in stress and pressure application.

The various weld types in this welding joint are:

- Square-groove weld or butt weld
- Bevel-groove weld
- V-groove weld
- J-groove weld
- U-groove weld
- Edge-flange weld
- Corner-flange weld

Welding Electrodes

A welder needs an electrode to generate an electric current to do arc welding. In welding, an electric current is conducted through an electrode which is used to join the parent metals. When you keep electrode tip near the parent metal electric current jumps from the electrode tip to the parent metal. The main purpose of electrodes used in welding is to create an electric arc. These electrodes can be positively charged anode or they can be negatively charged cathode.

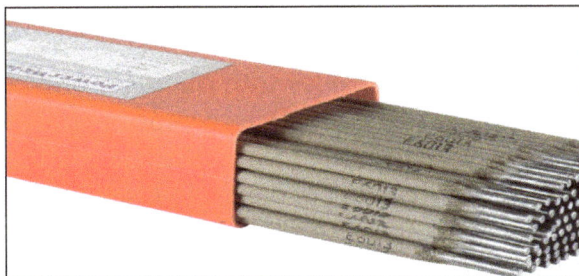

Factors that you must consider before selecting welding electrodes:

- The electrode rod should have greater tensile strength than the parent metals.
- You have to consider joint design, shape, specifications of base metals and welding positions.

Types of Welding Electrodes

Basically, depending upon the process there are two types of welding electrodes:

- Consumable Electrodes
- Non-Consumable Electrodes

Consumable Electrodes

Consumable electrodes have low melting point. These types of welding electrodes are preferred to use in Metal Inert Gas (MIG) welding. For making consumable electrodes, materials such as mild steel and nickel steel are used. The one precaution that you must take is to replace consumable electrodes after regular intervals. The only disadvantage of using such electrodes is that they don't have a large number of industry applications but at the same time they are easy to use and maintain.

Consumable electrodes are categorized as:

- Bare Electrodes
- Coated Electrodes

Bare Electrodes

Bare electrodes are electrodes without any type of coating and mostly used in applications where there is no need of coated electrode.

Coated Electrodes

Coated electrodes are classified according to the coating factor. Coating factor is the ratio of the diameter of the electrode to the diameter of the core wire.

So, following are sub types of coated electrodes:

- Light coated electrodes with coating factor of 1.25. Light coating applied to electrodes helps to remove impurities such as oxides and phosphorous. Light coating also helps in enhancing arc stability.
- Medium coated electrodes with coating factor of 1.45.
- Shielded arc or Heavily coated electrodes with coating factor ranging between 1.6 to 2.2. These electrodes have a proper and well defined composition. The heavily coated electrodes are designed in three types – electrodes with cellulose coating, electrodes with mineral coating and the electrodes with coating of both cellulose as well as mineral coating.

Non-Consumable Electrodes

These types of welding electrodes are also referred to as Refractory electrodes. There are again two sub-types of non-consumable electrodes:

- Carbon or Graphite electrodes: It is made up of carbon and graphite and mostly used in the applications of cutting and arc welding.

- Tungsten electrodes: Basically, it is consists of tungsten as the name itself suggests and it is a non-filler metal electrode.

As the name suggests, these types of welding electrodes are not consumed in the entire welding process or we can say more appropriately that they do not melt during welding. But practically, due to the vaporization and oxidation processes taking place during welding there is a little bit reduction in the length of the electrode. The non-consumable electrodes have high melting point and are unable to fill the gap in the workpiece. Non-consumable electrodes are made from materials such as pure tungsten, graphite or carbon coated with copper. The melting point of carbon is 3350 degree Celsius and that of Tungsten is 3422 degree Celsius. Non-consumable electrodes are used in Tungsten inert gas welding (TIG) and carbon arc welding.

Important Characteristics of Non-Consumable Electrodes

- While using non-consumable electrodes you have to use shielding gases. The shielding gases are inert gases and the reason behind to use them is to protect welding area from oxygen and surrounding atmosphere.

- The non-consumable electrodes are usually made as cathode and the workpieces as anode.

Classification of Welding Electrodes

This classification is based on practical approach towards using and selecting welding electrodes. The American Welding Society has classified electrodes in different formats for proper understanding of different electrodes easily and to identify them comfortably.

Suppose we Consider the Electrode named as E6018-X

- Here the E indicates that this is an electrode.

- The consecutive two digits after the letter E gives you the tensile strength of the electrode. This tensile strength is measured in psi and this strength is 1000 times the given number. That means here this tensile strength of the given electrode is 60,000 psi.

- Here 1 indicates the welding position. The welding position is indicated by 1,2 and 4.

  - 1 indicates flat, horizontal, vertical position.

  - 2 indicates flat, horizontal position.

  - 4 indicates flat, horizontal, vertically downward position.

- The number 8 gives you information about the type of coating and current used. It also tells about the penetration of electrode means the electrode may penetrate deep, low, medium.

- The X in the E6018-X tell us about the additional specifications of an electrode. Here the term X is not always mentioned. It is used only when an electrode has some additional features. This classification is applicable to mild steel coated electrode. If you consider other types of electrodes, the classification will be same but only the feature represented by X letter may vary.

- Some additional properties represented by letter X:

  1: It indicates that the electrode is more ductile and has high toughness.

  M: It is comfortable for military applications and low moisture content.

  H4, -H8,-H16: All represents maximum diffusible hydrogen limit measured in millimetres per 100 Grams. For instance, -H4 = 4 mL per 100 grams.

Precautions that you must take while Handling Welding Electrodes

- You must always keep electrodes dry.

- As moisture destroys electrode coating and is very harmful for electrodes. So, as soon as electrodes get dried, you must keep welding electrodes in moisture free environment. There are various containers available which give you moisture free experience.

Types of Welding Process

The different types of welding process are:

- Gas Welding:
  - Air acetylene
  - Oxy hydrogen welding
  - Oxy acetylene
- Arc Welding:
  - Shielded Metal Arc Welding (SMAW)
  - Gas Metal Arc Welding (GMAW) or (MIG)
  - Flux-Cored Arc Welding (FCAW)
  - Submerged Arc Welding (SAW)
  - Gas Tungusten Arc Welding (GTAW) or (TIG)
  - Plasma Arc welding (PAW)
  - Atomic Hydrogen Welding (AHW)
  - Carbon Arc Welding (CAW)
  - Electroslag Welding (ESW)
  - Electrogas Welding (EGW)

- Resistance Welding:

  ○ Seam welding

  ○ Projection welding'

  ○ Spot welding

  ○ Flash welding

  ○ Resistance Butt welding

- Radiant Energy Welding:

  ○ Electric or Electron beam welding

  ○ Laser beam welding

- Solid State Welding:

  ○ Cold Welding

  ○ Diffusion Welding

  ○ Explosive Welding

  ○ Forge Welding

  ○ Friction Welding

  ○ Hot Pressure Welding

  ○ Roll welding

  ○ Ultrasonic Welding

Basically there are three types of welding:

- Plastic Welding: It is also called as liquid-solid welding or pressure welding. Plastic welding is used in forge welding and resistance welding. This process includes combining small parts or pieces of metal. But before joining these small components they are strongly heated to achieve to a plastic state. After doing all these they are joined by applying external pressure.

- Cold Welding: It is popularly known as solid state welding which includes no application of heat but application of external pressure for the diffusion process. The parts are joined without contamination from atmospheric constituents. It is used in ultrasonic welding, friction welding and Explosive welding.

- Liquid State Welding: In this technique metal at the junction point is allow to solidify after heating. In this filler materials are also supplied to the metal. It's another name is Fusion welding. It involves gas welding, arc welding.

Gas Welding

Air Acetylene Welding

It is a type of welding process in which heat is produced by combining mixture of acetylene and air. In this process there is an increase in temperature of about 2700 degree C. There is formation of weld point without application of heat or without using filler metals. In this type of welding only one tank is used which makes it less expensive and easy to handle.

If we see the working of Bunsen burner that is totally based on this air acetylene welding.

Applications

- In case of leads having small area of cross section this method is used for welding.
- Also sheets of copper having small thickness are also welded by this process.

Oxy-Hydrogen Welding

In this process oxygen and hydrogen are used for fusing and cutting. The oxy-hydrogen flame produced in this process is typically of pale blue color. The temperature of this flame is about 2000 degree C. The workpiece is firstly heated and the melt bar fused in to make in the connection or joint. The time interval is not fixed for the completion of this process. It will vary from two minutes to five minutes depending on the size of sheets.

Disadvantages:

- There are more chances of leaking the gas in this process.
- It is more expensive than other alternative processes.

Uses:

- Melting of expensive metals.
- Polishing of acrylic glass surfaces.
- In metal manufacturing industries.

Oxy-Acetylene Welding

In 1903 Edmond and Charles Picard discover this welding type. This is one of the most important types of welding process commonly called as oxy-acetylene or oxy-fuel welding. The unique feature of this process is that it uses fuel gases such as hydrogen, natural gas, methane and oxygen to fuse and cut metal pieces. The flame temperature is about 3100 degree Celsius. It involves two processes:

Fusing or Welding

This process placed more emphasis on using welding torch to fuse metals. Tip of this torch is used to fuse metals. The torch consists of oxygen control unit, Fuel gas control valve and mixer.

Cutting or Slicing

Basically a cutting torch is used to crop the metal. Before cutting, metal components are heated up to ignition point by pressing oxygen blast lever present in the torch. Iron oxide and heat formation takes places after the reaction of metal with oxygen. The resultant heat is the deciding factor of the cutting process.

Uses:

- For polishing in glass companies.

- It is used in jewelry designing for the purpose of water welding,

- To obtain a bright light in theatres.

Precautions

- Less than 1/7 the capacity of the cylinder should be used per hour.

- Fuel leakage should be avoided.

- Cylinders should be properly monitored and maintained.

Resistance Welding

It is the types of welding Process mostly used in production sites for combining metal sheets and parts. The weld is made by passing a strong current through this junction to heat them and melt the metals. Factors affecting welding temperatures are electrode size, its specifications means geometry, current, and interval of welding time. Small pools of molten metal are produced at the point of most electrical resistance as an electric current is passed through the metal.

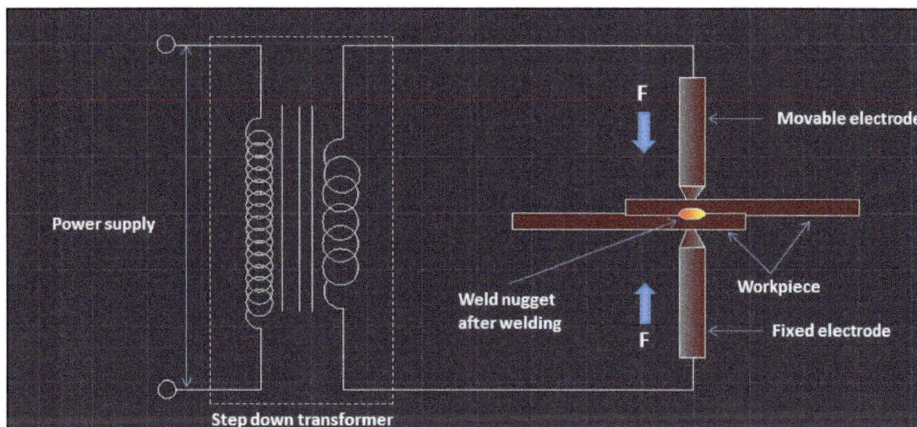

It mainly includes following processes:

Spot Welding

Spot types of welding is a resistance welding method implemented to fuse the two or more metal hangers, sheets. Basically, two copper electrodes are used to clamp the metal sheets and then current is passed through these sheets. When current is passed to the sheets through electrodes, gradually amount of heat starts generating as a result of strong electrical resistance at the contact point of electrodes and sheets. Due to rising temperature there is more dissipation of heat throughout the workpiece in just a second and then molten state grows to meet welding tips. So this is the process of spot welding.

Advantages:

- Filler materials are not required.

- It includes efficient energy use.

- Fast and easy automation.

Uses:

- It was used in making of BMW3 series.

- It is used in car industry to prepare a sheet of metal.

- It is also used in manufacturing process of batteries.

- It is also used in orthodontist's clinic.

Seam Welding

It is an automated types of welding technique and it produces an efficiently durable weld. It generally produces a weld at the connecting surfaces of two similar metals. It includes two electrodes of copper and mostly these electrodes are disc shaped and rotate as the material passes between them. The weld made from this process is stronger than material from which it is made. It was formerly used for the construction of beverage cans.

Flash Welding

Flash welding doesn't use any filler metals. There is a fix distance between metal pieces that are to be welded. On applying current to the metal, resistance is created between the spaces present among metals which generate required heat for the welding purpose.

Uses:

- It is used extensively in railroad industry.

- It is also used to fuse aluminium, copper and steel in various conductors.

Projection Welding

In these types of welding process, the weld is fixed to a particular position with the help of projections on the workpieces that we have to join. There is generation of heat at the projection in this process. Due to projection welding there is generation of internal cracks as well as ugly look of metal. You need not to take more safety measures while projection welding. Only simple goggles are required during this process.

Resistance Butt Welding

This technique involves application of heat and pressure to combine the two metal parts in localised area. It is generally used to make butt joints in wire as well as rods up to about 16 mm diameter. The larger and parallel surfaces of the pieces to be joined concentrate the strong heating at the interface. Current is passed through dies which in turn causes resistance heating of the weld area. This heating is larger in those areas where resistance is higher.

Advantages:

- It is superfast and clean process.

- It is better than flash welding due to its application over small components.

- There are some precautions should be taken during this welding process:

    ◦ Eyes should be protected by wearing safety helmets.

    ◦ Leather gloves should be used.

Arc Welding

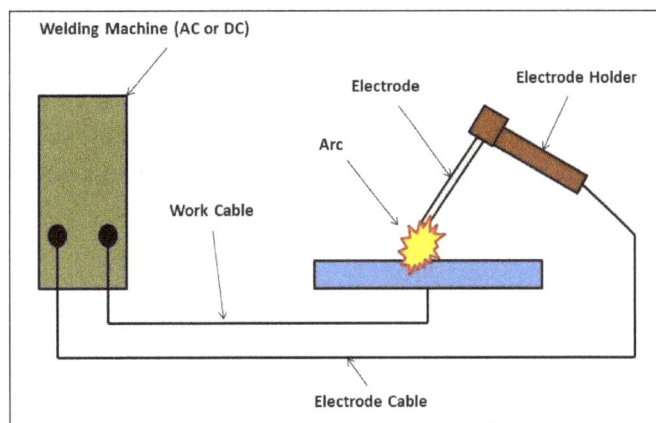

This process may be automated or manual. In these types of welding, metals are connected end to end with the help of electric supply which in turn results in sufficient amount of heat production for the welding purpose. This welded metal then allowed to solidify for their fusing. Direct current (DC) or alternating current (AC) can be used in this type. Mostly, constant current supply is used for this process. Sometimes, the zone or area is covered with vapor or slag. There are two types of electrodes which are used in this process:- 1) Consumable electrodes (2) Non-consumable electrodes

Shielded Metal Arc Welding

Shielded Metal Arc Welding is popularly called as Manual Metal Arc Welding (MMAW) or stick welding.Its process and structure is similar to that of arc welding as it is sub-type of arc welding. So as mentioned in arc welding in the same way an electric arc is created in this process. Electrode coated in core wire covered with flux. The temperature of about 6500 degree F is generated in this method. The molten metal is protected by gaseous oxides and nitrides with the gaseous shield so it is termed as Shielded Metal Arc Welding.

Uses:

- It is used in repairing of large machines.
- Usually used in pipeline welding.
- If proper safety measures are not taken then it is very dangerous process:
  - Welding helmets, long sleeve cloths and gloves should be wear during this process.
  - Advanced welding masks can save your life from toxic fumes.

Advantages:

- Cheap cost machinery.
- Works on rough surfaces.
- Machinery involve in the process is easy to handle.

Gas Tungsten Arc Welding (GTAW) or (TIG)

It is also called as Tungsten Inert Gas (TIG) welding. It uses non-consumable tungsten electrode for the welding purpose. Filler gas and shielding gases are used for protection.

To prevent contact between electrode and workpiece it is very necessary to keep a certain short arc length. An electric spark is created with the help of generator and the distance of 1.5-3 mm is maintained between electrode and workpiece. Holding filler rod close to an arc is very harmful as it results in melting of filler rod before making any contact with weld pool. At the end of the weld there is formation of solidified hole shaped or simply crater cracks.

Uses:

- This types of welding process is extensively used in aerospace welding.
- In the manufacturing of motorcycles and bikes.

Advantages:

- Higher deposition rates.
- Large and good quality of welds.
- Very thin materials can be easily welded.

Gas Metal Arc Welding (GMAW) or (MIG)

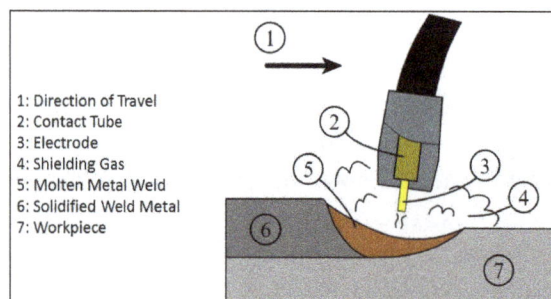

1: Direction of Travel
2: Contact Tube
3: Electrode
4: Shielding Gas
5: Molten Metal Weld
6: Solidified Weld Metal
7: Workpiece

It is mainly referred by its subtype Metal Inert Gas Welding (MIG). It is not a 100% manual process. It can be semi-automatic or fully automatic. It was originally implemented to aluminium as well as non-ferrous metals but soon it is applied to steels because of its quick welding time compared to similar processes. In this process, welder have to use one hand for holding torch and there is a filler wire in the other hand. This is a very simple and easy technique of welding and generally require a span of one week for learning it. Its working procedure is almost similar to that of arc welding. The shielding gases are carbon dioxide or the mixture of carbon dioxide and argon-carbon dioxide. It is useful in fabrication and automobile industry.

Advantages:

- Reduce heat inputs

- Easy to learn and use

- Efficient electrodes

Flux Cored Arc Welding

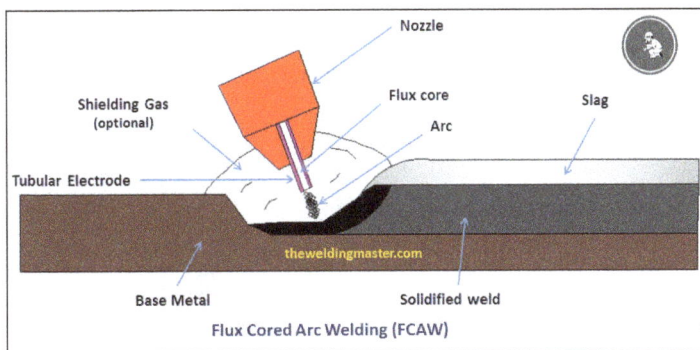

Flux Cored Arc Welding (FCAW)

It is similar to GMAW only it requires continuously used fed electrode and uses tabular wire filled with flux. There are two types of FCAW -one uses shielding gas while other don't require any shielding gas. FCAW with no shielding gas is made by tabular consumable electrode containing flux core. It is attractive due to its portability and good penetrating power into metal. Another type of welding which uses shielding gas is termed as dual shield welding. This FCAW was purposefully developed for welding thicker metals and structural steels. The biggest advantage of this type of FCAW is that it produces high quality welds than SMAW or GMAW.

Advantages:

- It is a perfect all position process.

- Less skills required compared to SMAW.

- No requirement of shielded gas.

Uses:

- Used in quick automotive machines

- Welding of stainless steel and nickel alloys.

Submerged Arc Welding

In SAW types of welding Process, there is a prevention of welded zone from atmospheric contamination by submerging under a flux which contains lime, silica and calcium fluoride. The molten conductive flux provides a current path between the electrode and the workpiece. In market there are semi-automatic guns containing pressurised flux are available. The arc length is kept constant by self-adjusting arc principle. Arc length is inversely proportional to the voltage increase.

Uses:

- In corrosion resistant layer of steel.

- In vessel construction it is mostly used.

- The welding process containing alloys of nickel and low metals.

Advantages:

- More than half of the flux can be reuse and recycle.

- The welding process is very deep.

- Production of welds which are ductile and resistant to corrosion.

Plasma Arc Welding

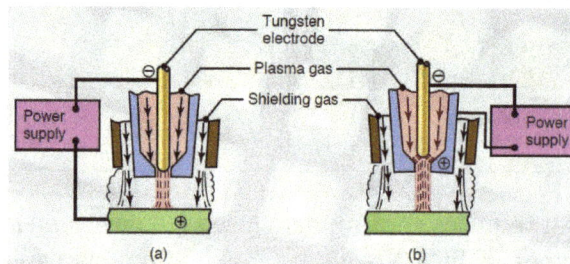

Helium is in process when there is requirement of broad heat input pattern. There are at least two separate gas flows are needed in PAW.

Some plasma processes are:

- Plasma arc cutting

- Plasma arc spraying

- Plasma arc surfacing

Atomic Hydrogen Welding

In the presence of shielding environment of hydrogen, arc is applied between two tungsten electrodes in AHW. Due to this arc there is an increment in temperature up to 4000 degree C.

The arc is used well apart from the workpiece. By changing the distance between arc and the workpiece the energy is controlled in AHW. Now a days gas metal arc welding is used rather than

HW due to availability of cheap inert gas. The filler metal may be used in this process. This method is also called as arc-atom welding. The torch used is atomic hydrogen torch or Langmuir torch.

Carbon Arc Welding

In CAW types of welding technique, an arc is heat between carbon electrode and workpiece to produce merging of metals. This arc produces a temperature of about 3000 degree C. At this temperature separate metal forms a bond and welding takes place. GTAW, PAW, AHAW have replaced CAW in this modern world. There are also 2 subtypes in CAW:

- TCAW (Twin Carbon Arc Welding)

    ◦ In Twin Carbon Arc Welding arc is placed between 2 carbon electrodes

- CAW-G (Gas Carbon Arc Welding)

    ◦ It is Gas Carbon Arc Welding. Generally a filler metal is used for the bondage in workpieces.

Electroslag Welding

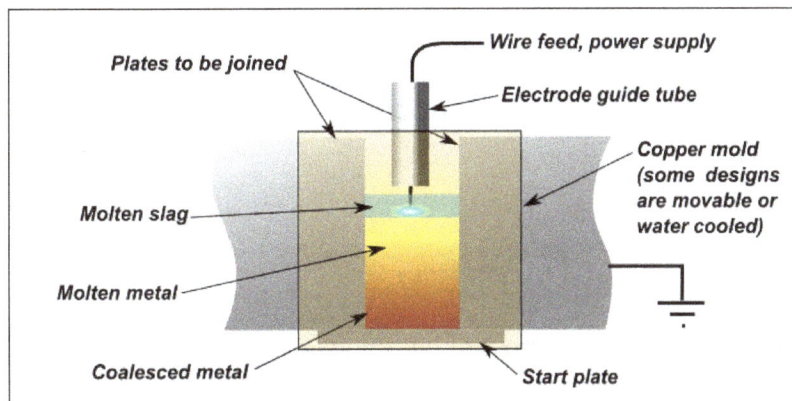

Mostly, electroslag types of welding are used for welding of thick materials. Due to the dissimilarity in arc positions there exist a difference between electroslag and electrogas welding process. An electric arc is hit at the weld location and additional flux is added until the slag extinguishes the arc. It is used to combine steel sections that are very thick. It is also implemented on structural steel plates. It requires current of about 600 A and voltage of about 45-55 V.

Advantages:

- Very large deposition rate of about 20 kg per hour.

- Safe and efficient process.

- Less requirement of labour work of high skills.

Electrogas Welding

There is no use of shielding gas and also no pressure is applied. The only difference between EGS and EGW is that arc in EGW is not quenched. It is implemented to most of steels including low and medium steels. It requires the current ranging from 100 to 800 A and voltage ranging from 30-40 volt. Carbon dioxide is widely used as shielding gas. Large amount of molten metal is used in this types of welding process which increases the safety measures that have to be taken care properly. It is advisable to wear a helmet during this process.

Uses:

- In shipbuilding processes

- In the development of storage tanks.

Stud Arc Welding

It is the most extensively used fastening process. It can weld any type of metal stud to work piece. There is requirement of DC power supply, a welding gun and metal fasteners. There is no much pressure requirement in this method.

Advantages:

- It is used in jewelry production.

- It is used in pumps, motors and electronic gadgets.

- Used in household utensils and hardwares.

Radiant Energy Welding

In radiant energy welding we uses radiant energy produced for the welding of two metal pieces together.

- Laser Beam Welding: It is a process of joining two or more pieces of metal with the help of laser beam. The weld of this type is formed as a result of light beam falling on the metal. This process takes place in few seconds. As a laser beam is monochromatic it will produce a large amount of heat after hitting the metal surface. This laser beam contains highly concentrated energy due to its coherency and single frequency.

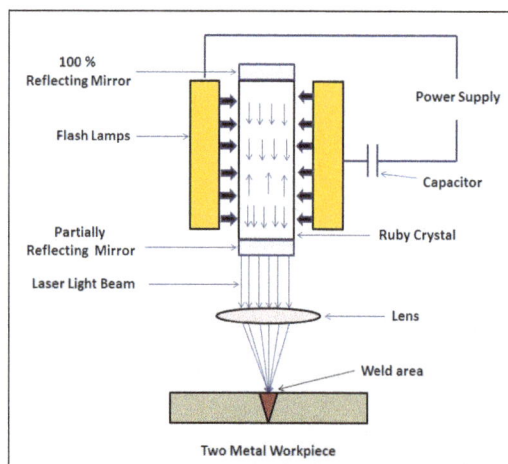

For welding purpose there is need of sharply concentrated rays. The size of the laser spot can be small as 0.076 mm. The lasers which are used for cutting purpose in these types of welding have very high cost.

Advantages:

- Easy to use in open environment.

- Highly efficient in welding small components.

- There is no need of contact with any material.

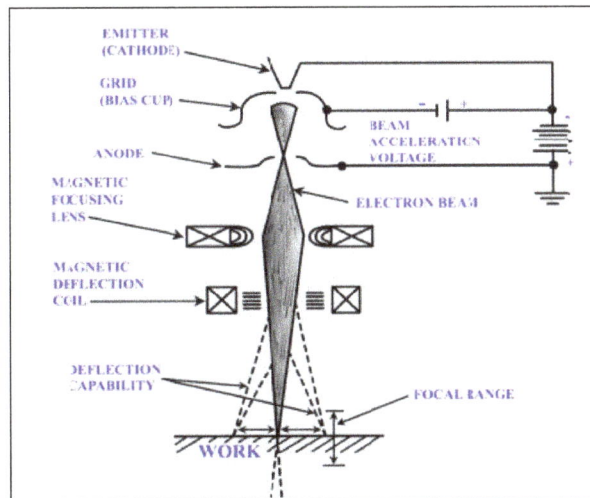

- Electron Beam Welding (EBW): EBW is an advanced types of welding technique in which a beam of highly energetic electrons is focused on the metals to be combined. In this process kinetic energy of these highly accelerated electrons is converted into heat energy. Due to this reason the workpieces melt and flow upon incident of electron beam.

The machine uses for this purpose consist of four compartments:

- Electron gun: Generating highly accelerated electrons.

- Vacuum chamber

- Control and power supply system.

- Electric and magnetic field generator.

Solid State Welding

It is a collection of welding methods that produces merging at temperature especially below the melting point of base metals that are to be joined. The Various types of welding process that come under solid state welding are described below:

Cold Welding

- This Welding technique uses pressure at room temperature to produce coalescence of metals.

Welding is achieved by applying high pressure on clean materials. Cold welding is usually implemented to combine the ductile metals. For instance, aluminium and copper can be joined together by cold welding.

Forge Welding

In this technique there is permanent deformation at the interface due to application of sufficient pressure and resulting in coalescence of metals by heating them. It was formerly called as hammer welding. This process is of minor use in industry today.

Uses:

- It is used in the manufacturing of prison cells and shotgun barrels.

- Also in the manufacturing processes of swords and cookware.

Diffusion Welding

Diffusion welding is solid state types of welding which is capable of joining similar as well as dissimilar materials. It involves no filler metal. If the metal has strong oxide layers then diffusion welding must be done in vaccum or in the presence of inert gas conditions.

Uses:

- It is used in electronics and nuclear industry.

- Metals like titanium, zirconium can be easily melted with the help of this process.

- In military aircrafts there is need of less expensive metals due to this process.

Advantages:

- Tensile testing can be easily done.

- Easy processing of similar as well as dissimilar metals.

Explosive Welding (EXW)

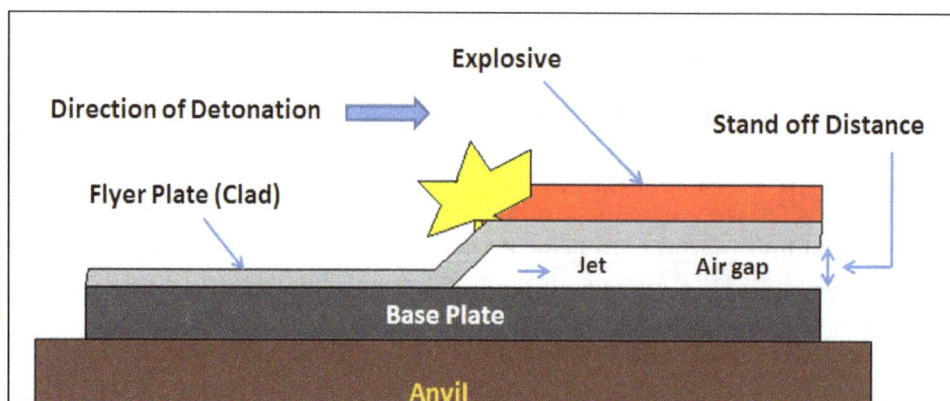

EXW was developed during world war 2. It is mostly based on metallurgical bonding. With the help of chemical explosives one of the parts is accelerated. These method is used to develop plates and tube sheets. It is used in small shops for joining cooling fan.

Advantages:

- Easy bonding between two metals that cannot be welded by other alternatives.
- Quick welding of large areas.
- Welding is efficient and clean.

Friction Welding

In FRW there is heat generation due to friction between workpieces. Friction surfacing is the procedure involved in FRW where a substrate is protected with coating substance. It is used with thermoplastics.

Advantages:

- They are melt free.
- Relative motion between workpieces cleans their surfaces.
- In aerospace it is useful to join dissimilar materials like aluminium with steels.

Hot Pressure Welding

It is the subtype of solid state types of welding process in which welds are produced between two connecting metals. There is generation of plastic deformation and there is no requirement of filler material in this technique. Pressure of 40-70 MPa is applied during the whole process. Heat may be generated with the help of eddy currents.

Advantages:

- Less expensive machinery.
- Quality and quick weld production.

Roll Welding

It is the resultant of flat rolling of metal sheets. The two or more metal pieces passed between two rollers and there is large pressure applied on rollers to reduce the effective metal thickness. To improve quality and strength of weld it should be pre-heated before welding. It should be performed in quite warm conditions.

Uses:

- In refrigerators there are heat exchangers which are created from this technique.
- Used in efficient cladding of metal sheets.

Ultrasonic Welding

It is the most advanced types of welding method, in which high frequency ultrasonic vibrations are applied to workpieces for the formation of weld. It is widely used for plastics and combining dissimilar materials. The frequency of ultrasonic rays varies from 15kHz-70kHz.

Uses:

- Combining thermoplastics.

- Used in motors, capacitors, transformers.

- In microcircuits, this is the most dominant welding technology.

Welding Hazards

Health Hazards of Welding

Gases and Fumes

Welding "smoke" is a mixture of very fine particles (fumes) and gases. Many of the substances in welding smoke, such as chromium, nickel, arsenic, asbestos, manganese, silica, beryllium, cadmium, nitrogen oxides, phosgene, acrolein, fluorine compounds, carbon monoxide, cobalt, copper, lead, ozone, selenium and zinc, can be extremely toxic.

Generally, welding fumes and gases come from:

- The base material being welded or the filler material that is used;

- Coatings and paints on the metal being welded, or coatings covering the electrode;

- Shielding gases supplied from cylinders;

- Chemical reactions which result by the action of ultraviolet light from the arc and heat;

- Process and consumables used; and

- Contaminants in the air, for example vapors from cleaners and degreasers.

The health effects of welding exposures are difficult to list, because the fumes may contain so many different substances that are known to be harmful. The individual components of welding smoke can affect any part of the body, including the lungs, heart, kidneys and central nervous system. Welders who smoke may be at greater risk of health impairment than welders who do not smoke, although all welders are at risk. 2 Exposure to welding smoke may have short-term and long-term health effects. These effects are described below:

Short-Term (Acute) Health Effects

Exposure to metal fumes (such as zinc, magnesium, copper, and copper oxide) can cause metal fume fever. Symptoms of metal fume fever may occur 4 to 12 hours after exposure, and include chills, thirst, fever, muscle ache, chest soreness, coughing, wheezing, fatigue, nausea and a metallic taste in the mouth. Welding smoke can also irritate the eyes, nose, chest, and respiratory tract, and cause coughing, wheezing, shortness of breath, bronchitis, pulmonary edema (fluid in the lungs) and pneumonitis (inflammation of the lungs). Gastrointestinal effects, such as nausea, loss of appetite, vomiting, cramps, and slow digestion, have also been associated with welding. Some components of welding fume, for example cadmium, can be fatal in a short time. Gases given off by the welding process can also be extremely dangerous. For example, ultraviolet radiation given off by welding reacts with oxygen and nitrogen in the air to form ozone and nitrogen oxides. These gases are deadly at high doses and can also cause irritation of the nose and throat and serious lung disease. Ultraviolet rays given off by welding can react with chlorinated hydrocarbon solvents, such as 1, 1, 1-trichloroethane, trichloroethylene, methylene chloride, and perchloroethylene, to form phosgene gas. Even a very small amount of phosgene may be deadly, although early symptoms of exposure -- dizziness, chills, and cough -- usually take 5 or 6 hours to appear. Arc welding should never be performed within 200 feet of degreasing equipment or solvents.

Long-Term (Chronic) Health Effects

Studies of welders, flame cutters, and burners have shown that welders have an increased risk of lung cancer, and, possibly cancer of the larynx (voice box) and urinary tract.

These findings are not surprising in view of the large quantity of toxic substances in welding smoke, including cancer-causing agents such as cadmium, nickel, beryllium, chromium, and arsenic.

Welders may also experience a variety of chronic respiratory (lung) problems, including bronchitis, asthma, pneumonia, emphysema, pneumoconiosis (refers to dust-related diseases), decreased lung capacity, silicosis (caused by silica exposure) and siderosis (a dustrelated disease caused by iron oxide dust in the lungs).

Other health problems that appear to be related to welding include: heart disease, skin diseases, hearing loss and chronic gastritis (inflammation of the stomach), gastroduodenitis (inflammation of the stomach and small intestine) and ulcers of the stomach and small intestine. Welders exposed to heavy metals such as chromium and nickel have also experienced kidney damage.

Welding also poses reproductive risks to welders. A recent study found that welders, and especially welders who worked with stainless steel, had poorer sperm quality than men in other types of work. Several studies have shown an increase in either miscarriages or delayed conception among welders or their spouses. Possible causes include exposure to: (1) metals, such as aluminium, chromium, nickel, cadmium, iron, manganese, and copper, (2) gases, such as nitrous gases and ozone, (3) heat and (4) ionizing radiation (used to check the welding seams). Welders who perform welding or cutting on surfaces covered with asbestos insulation are at risk of asbestosis, lung cancer, mesothelioma and other asbestos-related diseases. Employees should be trained and provided with the proper equipment before welding near materials that contain asbestos.

Other Health Hazards

Heat

The intense heat of welding and sparks can cause burns. Eye injuries have resulted from contact with hot slag, metal chips, sparks, and hot electrodes.

In addition, excessive exposure to heat can result in heat stress or heat stroke. Welders should be aware of the symptoms, such as fatigue, dizziness, loss of appetite, nausea, abdominal pain, and irritability. Ventilation, shielding, rest breaks, and staying hydrated will protect against heat related hazards.

Visible Light and Ultraviolet and Infrared Radiation

The intense light associated with arc welding can cause damage to the retina of the eye, while infrared radiation may damage the cornea and result in the formation of cataracts.

Invisible ultraviolet light (UV) from the arc can cause "arc eye" or "welder's flash" after even a brief exposure (less than one minute). The symptoms of arc eye usually occur many hours after exposure to UV light, and include a feeling of sand or grit in the eye, blurred vision, intense pain, tearing, burning, and headache.

The arc can reflect off surrounding materials and burn co-workers who work nearby. About half of welder's flash injuries occur in co-workers who are not welding. Welders and cutters who continually work around ultraviolet radiation without proper protection can suffer permanent eye damage.

Exposure to ultraviolet light can also cause skin burns similar to sunburn and increase the welder's risk of skin cancer.

Noise

Exposure to loud noise can permanently damage welders' hearing. Noise also causes stress and increased blood pressure, and may contribute to heart disease. Working in a noisy environment for long periods of time can make workers tired, nervous, and irritable. If you work in a noisy area, the OSHA Noise Standard requires your employer to test for noise levels to determine your exposure. If your noise exposure exceeds 85 decibels averaged over 8 hours, your employer must provide you with free hearing protection and annual hearing tests.

Musculoskeletal Injuries

Welders have a high prevalence of musculoskeletal complaints, including back injuries, shoulder pain, tendinitis, reduced muscle strength, carpal tunnel syndrome, white finger, and knee joint diseases. Work postures (especially welding overhead, vibration, and heavy lifting) can all contribute to these disorders. These problems can be prevented by:

- Proper lifting;
- Not working in one position for long periods of time;
- Keeping the work at a comfortable height;
- Using a foot rest when standing for long periods;
- locating tools and materials conveniently; and
- minimizing vibration.

Safety Hazards of Welding

Electrical Hazards

Even though welding generally uses low voltage, there is still a danger of electric shock. The environmental conditions of the welder (such as wet or cramped spaces) may make the likelihood of a shock greater. Falls and other accidents can result from even a small shock; brain damage and death can result from a large shock.

Dry gloves should always be worn to protect against electric shock. The welder should also wear rubber-soled shoes and use an insulating layer, such as a dry board or a rubber mat, for protection on surfaces that can conduct electricity.

The piece being welded and the frames of all electrically powered machines must be grounded. The insulation on electrode holders and electrical cables should be kept dry and in good condition. Electrodes should not be changed with bare hands, with wet gloves, or when standing on wet floors or grounded surfaces.

Fires and Explosions

The intense heat and sparks produced by welding, or the welding flame, can cause fires or explosions if combustible or flammable materials are nearby.

Welding or cutting should be performed only in areas that are free of combustible materials, including trash, wood, paper, textiles, plastics, chemicals and flammable dusts, liquids and gases (vapors can travel several hundred feet). Those that cannot be removed should be covered with a tight-fitting flame-resistant material. Doorways, windows, cracks and other openings should be covered.

Never weld on containers that have held a flammable or combustible material unless the container is thoroughly cleaned or filled with an inert (non-reactive) gas. Explosions, fires, or release of toxic vapors may result. Containers with unknown contents should be assumed to be flammable or combustible.

A fire inspection should be performed before leaving the work area and for at least 30 minutes after the operation is completed. Fires extinguishers should be nearby.

Dangerous Machinery

All machines with moving parts must be guarded to prevent workers' hair, fingers, clothing, etc. from getting caught.

When repairing machinery by welding or brazing, power must be disconnected, locked out, and tagged so that the machinery cannot start up accidentally.

Trips and Falls

To prevent trips and falls, keep welding areas clear of equipment, machines, cables and hoses, and use safety lines or rails.

Hazards of Welding in Confined Spaces

A confined space is a small or crowded area with limited access and little or no air flow or ventilation. Adequate ventilation is essential for working in confined spaces. Dangerous concentrations of toxic fumes and gases can build up very quickly in a small space. Unconsciousness or death from suffocation can occur rapidly because welding processes can use up or displace oxygen in the air. High concentrations of some fumes and gases can also be very explosive.

All workers who may enter dangerous areas, either on a regular basis or in an emergency situation, should be trained on rescue procedures, personal protective equipment, use of safety equipment, and proper procedures for entering and exiting a confined space.

- The worker inside the confined space should be equipped with a safety harness, a lifeline, and appropriate personal protective clothing, including a self-contained breathing apparatus. (Never use an air purifying respirator.)

- Gas cylinders and welding power sources should be located in a secure position outside of the confined space.

- A trained helper must be stationed outside of the confined space and equipped with appropriate gear (including a fire extinguisher and personal protective equipment), to assist or rescue the worker inside the confined space if necessary. If the standby person notices any indications of intoxication or decreased alertness from the "inside" worker, the inside worker should be removed from the area immediately.

- All confined spaces should be tested before entering for toxic, flammable, or explosive gases or vapors, and oxygen level. Continuous air monitoring may be necessary during welding. No worker should enter a confined space where the percentage of oxygen is below 19.5 percent unless he or she is equipped with a supplied-air respirator.

- Never use oxygen for ventilation.

- Use continuous mechanical ventilation and a respirator whenever you weld or perform thermal cutting in a confined space.

- All pipes, ducts, and power lines connected to the space, but not necessary to the operation, should be disconnected or shut off. All shutoff valves and switches should be tagged and locked out so they cannot be restarted accidently.

- All unnecessary torches and other gas or oxygen-supplied equipment should be removed from the confined space.

Hazards of Compressed Gases

Gas welding and flame cutting use a fuel gas and oxygen to produce heat for welding. For high-pressure gas welding, both the oxygen and the fuel gas (acetylene, hydrogen, propane, etc.) supplied to the torch are stored in cylinders at high pressure.

The use of compressed-gas cylinders poses some unique hazards to the welder. Acetylene is very explosive. It should be used only with adequate ventilation and a leak detection program.

Oxygen alone will not burn or explode. At high oxygen concentrations, however, many materials (even those that are difficult to burn in air, such as normal dust, grease, or oil) will burn or explode easily.

- All cylinders should have caps or regulators.

- Only pressure regulators designed for the gas in use should be fitted to cylinders.

- Compressed gas cylinders, all pressure relief valves and all lines should be checked before and during welding operations.

- Blowpipes must be kept in good condition and cleaned at regular intervals.

- Hoses and fittings should be kept in good condition and checked regularly.

- Cylinders must be stored upright so that they will not fall over.

- Oxygen and fuel cylinders must be stored separately, away from heat and sunlight, and only in a dry, well-ventilated, fire-resistant area that is at least 20 feet away from flammable materials such as paint, oil, or solvents.

- Be aware of backfires and flashbacks. These are danger signals and should prompt immediate corrective action.

- Close cylinder valves when work is finished. Put valve protection caps in place and release pressure in regulators and hose lines before cylinders are moved or placed in storage.

2
Gas Welding

The process of melting and joining metals by using flame caused by the reaction of fuel gas and oxygen is called gas welding. It can be categorized into oxy fuel welding and oxy hydrogen welding. The topics elaborated in this chapter will help in gaining a better perspective about these types of gas welding as well as welding flames.

Gas welding is the most important type of welding process. It is done by burning of fuel gases with the help of oxygen which forms a concentrated flame of high temperature. This flame directly strikes the weld area and melts the weld surface and filler material. The melted part of welding plates diffused in one another and create a weld joint after cooling. This welding method can be used to join most of common metals used in daily life.

Equipment

Welding Torch

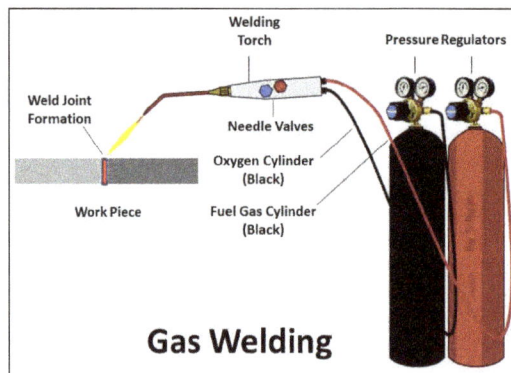

Gas Welding

Welding torches are most important part of gas welding. Both the fuel gas and oxygen at suitable pressure fed through hoses to the welding torch. There are valves for each gas witch control the flow of gases inside the torch. Both gases mixed there and form a flammable mixture. These gases ignite to burn at the nozzle. The fire flame flow through nozzle and strikes at welding plates. The nozzle thickness depends on the size of the welding plates and material to be welded.

Oxygen Cylinder

For proper burning of fuel, appropriate amount of oxygen required. This oxygen supplied by a oxygen cylinder. A black line is used to indicate oxygen cylinder.

Gas cylinder is filled either by oxy acetylene gas, hydrogen gas, natural gas or other flammable gas. The fuel gas selection is depends on the welding material. Mostly oxy acetylene gas is used for all general purpose of welding. Normally these cylinders have Maroon line to indicate it. The fuel gases passes through it.

Pressure Regulator

Both oxygen and fuel gases are filled in cylinder at high pressure. These gases cannot use at this high pressure for welding work so a pressure regulator is used between flow. It supplies oxygen at pressure about 70 – 130 KN / M2 and gas at 7 – 103 KN / M2 to the welding torch.

Goggles and Gloves

These are use for safety purpose of welder. It protects eyes and hand from radiation and flame of fire.

Working

Gas welding process is quite simpler compare to arc welding. In this process all the equipment are connected carefully. The gas cylinder and oxygen cylinder connected to the welding torch through pressure regulators. Now the regulate pressure of gas and oxygen supplied to the torch where they properly mixed. The flame is ignited by a striker. Take care the tip of torch is pointing downward. Now the flame is controlled through valves situated in welding torch. The flame is set at natural flame or carburizing flame or oxidizing flame according to the welding condition. Now the welding torch moved along the line where joint to be created. This will melt the interface part and join them permanently.

Application

- It is used to join thin metal plates.

- It can used to join both ferrous and non-ferrous metals.

- Gas welding mostly used in fabrication of sheet metal.

- It is widely used in automobile and aircraft industries.

Advantages and Disadvantages

Advantages:

- It is easy to operate and does not required high skill operator.

- Equipment cost is low compare to other welding processes like MIG, TIG etc.

- It can be used at site.

- Equipment's are more portable than other type of welding.

- It can also be used as gas cutting.

Disadvantages:

- It provides low surface finish. This process needs a finishing operation after welding.

- Gas welding have large heat affected zone which can cause change in mechanical properties of parent material.

- Higher safety issue due to naked flame of high temperature.

- It is Suitable only for soft and thin sheets.

- Slow metal joining rate.

- No shielding area which causes more welding defects.

Types of Welding Flames

There are three basic flame types: neutral (balanced), excess acetylene (carburizing), and excess oxygen (oxidizing) as shown below. A neutral flame is named neutral since in most cases will have no chemical effect on the metal being welded. A carburizing flame will produce iron carbide, causing a chemical change in steel and iron. For this reason a carburizing flame is not used on metals that absorb carbon. An oxidizing flame is hotter than a neutral flame and is often used on copper and zinc.

Welding Torch Flame Types

Neutral Welding Flame

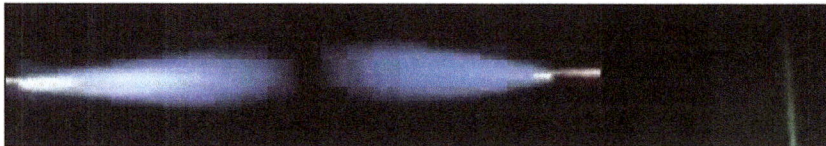

Carburizing flame (left), Neutral flame (center), Oxidizing flame (right).

The neutral flame has a one-to-one ratio of acetylene and oxygen. It obtains additional oxygen from the air and provides complete combustion. It is generally preferred for welding. The neutral flame has a clear, well-defined, or luminous cone indicating that combustion is complete.

Neutral welding flames are commonly used to weld:

- Mild steel

- Stainless steel

- Cast Iron

- Copper

- Aluminium

The welding flame should be adjusted to neutral before either the carburizing or oxidizing flame mixture is set. There are two clearly defined zones in the neutral flame. The inner zone consists of a luminous cone that is bluish-white. Surrounding this is a light blue flame envelope or sheath. This neutral flame is obtained by starting with an excess acetylene flame in which there is a "feather" extension of the inner cone. When the flow of acetylene is decreased or the flow of oxygen increased the feather will tend to disappear. The neutral flame begins when the feather disappears.

The neutral or balanced flame is obtained when the mixed torch gas consists of approximately one volume of oxygen and one volume of acetylene. It is obtained by gradually opening the oxygen valve to shorten the acetylene flame until a clearly defined inner cone is visible. For a strictly neutral flame, no whitish streamers should be present at the end of the cone. In some cases, it is desirable to leave a slight acetylene streamer or "feather" 1/16 to 1/8 in. (1.6 to 3.2 mm) long at the end of the cone to ensure that the flame is not oxidizing. This flame adjustment is used for most welding operations and for preheating during cutting operations. When welding steel with this flame, the molten metal puddle is quiet and clear. The metal flows easily without boiling, foaming, or sparking.

In the neutral flame, the temperature at the inner cone tip is approximately 5850 °F (3232 °C), while at the end of the outer sheath or envelope the temperature drops to approximately 2300 °F (1260 °C). This variation within the flame permits some temperature control when making a weld. The position of the flame to the molten puddle can be changed, and the heat controlled in this manner.

Carburizing Flame

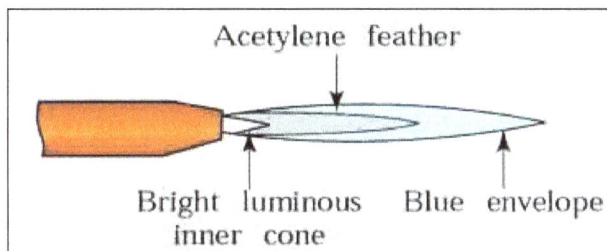

Components of a Carburizing Welding Flame.

The carburizing flame has excess acetylene, the inner cone has a feathery edge extending beyond it. This white feather is called the acetylene feather. If the acetylene feather is twice as long as the inner cone it is known as a 2X flame, which is a way of expressing the amount of excess acetylene. The carburizing flame may add carbon to the weld metal.

Reducing or carburizing welding flames are obtained when slightly less than one volume of oxygen is mixed with one volume of acetylene. This flame is obtained by first adjusting to neutral and then slowly opening the acetylene valve until an acetylene streamer or "feather" is at the end of the inner cone. The length of this excess streamer indicates the degree of flame carburization. For most welding operations, this streamer should be no more than half the length of the inner cone.

The reducing or carburizing flame can always be recognized by the presence of three distinct flame zones. There is a clearly defined bluish-white inner cone, white intermediate cone indicating the amount of excess acetylene, and a light blue outer flare envelope. This type of flare burns with a coarse rushing sound. It has a temperature of approximately 5700 °F (3149 °C) at the inner cone tips.

When a strongly carburizing flame is used for welding, the metal boils and is not clear. The steel, which is absorbing carbon from the flame, gives off heat. This causes the metal to boil. When cold, the weld has the properties of high carbon steel, being brittle and subject to cracking.

A slight feather flame of acetylene is sometimes used for back-hand welding. A carburizing flame is advantageous for welding high carbon steel and hard facing such nonferrous alloys as nickel and Monel. When used in silver solder and soft solder operations, only the intermediate and outer flame cones are used. They impart a low temperature soaking heat to the parts being soldered.

Oxidizing Flame

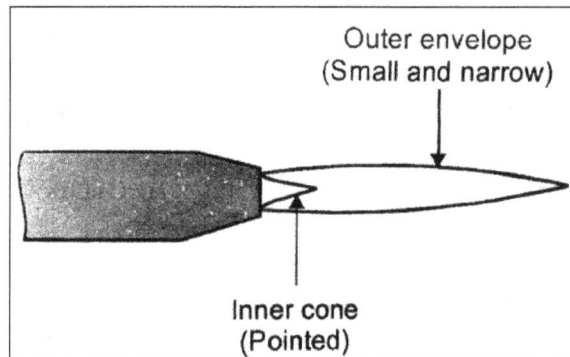

Components of an Oxidizing Welding Flame.

Oxidizing welding flames are produced when slightly more than one volume of oxygen is mixed with one volume of acetylene. To obtain this type of flame, the torch should first be adjusted to a neutral flame. The flow of oxygen is then increased until the inner cone is shortened to about one-tenth of its original length. When the flame is properly adjusted, the inner cone is pointed and slightly purple. An oxidizing flame can also be recognized by its distinct hissing sound. The temperature of this flame is approximately 6300°F (3482 °C) at the inner cone tip.

Oxidizing welding flames are commonly used to weld these metals:

- Zinc

- Copper

- Maganese steel

- Cast iron

When applied to steel, an oxidizing flame causes the molten metal to foam and give off sparks. This indicates that the excess oxygen is combining with the steel and burning it. An oxidizing flame should not be used for welding steel because the deposited metal will be porous, oxidized, and brittle. This flame will ruin most metals and should be avoided.

A slightly oxidizing flame is used in torch brazing of steel and cast iron. A stronger oxidizing flame is used in the welding of brass or bronze.

In most cases, the amount of excess oxygen used in this flame must be determined by observing the action of the flame on the molten metal.

Mapp Gas Welding Flames

The heat transfer properties of primary and secondary flames differ for different fuel gases. MAPP gas has a high heat release in the primary flame, and a high heat release in the secondary. Propylene is intermediate between propane and MAPP gas. Heating values of fuel gases.

The coupling distance between the work and the flame is not nearly as critical with MAPP gas as it is with other fuels.

Adjusting a MAPP gas flame. Flame adjustment is the most important factor for successful welding or brazing with MAPP gas. As with any other fuel gas, there are three basic MAPP gas flames: carburizing, neutral, and oxidizing.

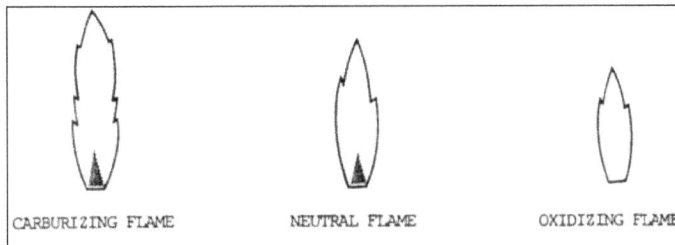

CARBURIZING FLAME   NEUTRAL FLAME   OXIDIZING FLAME

A carburizing flame looks much the same with MAPP gas or acetylene. It has a yellow feather on the end of the primary cone. Carburizing flames are obtained with MAPP gas when oxyfuel ratios are around 2.2:1 or lower. Slightly carburizing or "reducing" flames are used to weld or braze easily oxidized alloys such as aluminium.

As oxygen is increased, or the fuel is turned down, the carburizing feather pulls off and disappears. When the feather disappears, the oxyfuel ratio is about 2.3:1. The inner flame is a very deep blue. This is the neutral MAPP gas flame for welding. The flame remains neutral up to about 2.5:1 oxygen-to-fuel ratio.

Increasing the oxygen flame produces a lighter blue flame, a longer inner cone, and a louder burning sound. This is an oxidizing MAPP gas flare. An operator experience with acetylene will immediately adjust the MAPP gas flame to look like the short, intense blue flame typical of the neutral acetylene flame setting. What will be produced, however, is a typical oxidizing MAPP gas flame. With certain exceptions such as welding or brazing copper and copper alloys, an oxidizing flame is the worst possible flame setting, whatever the fuel gas used. The neutral flame is the principle setting for welding or brazing steel. A neutral MAPP gas flame has a primary flame cone abut 1-1/2 to 2 times as long as the primary acetylene flame cone.

Oxy-Fuel Welding and Cutting

Oxy-fuel welding (commonly called oxyacetylene welding, oxy welding, or gas welding in the U.S.) and oxy-fuel cutting are processes that use fuel gases (or liquid fuels such as gasoline) and oxygen to weld or cut metals. French engineers Edmond Fouché and Charles Picard became the first to develop oxygen-acetylene welding in 1903. Pure oxygen, instead of air, is used to increase the flame temperature to allow localized melting of the workpiece material (e.g. steel) in a room environment. A common propane/air flame burns at about 2,250 K (1,980 °C; 3,590 °F), a propane/oxygen flame burns at about 2,526 K (2,253 °C; 4,087 °F), an oxyhydrogen flame burns at 3,073 K (2,800 °C; 5,072 °F) and an acetylene/oxygen flame burns at about 3,773 K (3,500 °C; 6,332 °F).

During the early 20th century, before the development and availability of coated arc welding electrodes in the late 1920s that were capable of making sound welds in steel, oxy-acetylene welding was the only process capable of making welds of exceptionally high quality in virtually all metals in commercial use at the time. These included not only carbon steel but also alloy steels, cast iron, aluminium, and magnesium. In recent decades it has been superseded in almost all industrial uses by various arc welding methods offering greater speed and, in the case of gas tungsten arc welding, the capability of welding very reactive metals such as titanium. Oxy-acetylene welding is still used for metal-based artwork and in smaller home-based shops, as well as situations where accessing electricity (e.g., via an extension cord or portable generator) would present difficulties. The oxy-acetylene (and other oxy-fuel gas mixtures) welding torch remains a mainstay heat source for manual brazing and braze welding, as well as metal forming, preparation, and localized heat treating. In addition, oxy-fuel cutting is still widely used, both in heavy industry and light industrial and repair operations.

In oxy-fuel welding, a welding torch is used to weld metals. Welding metal results when two pieces are heated to a temperature that produces a shared pool of molten metal. The molten pool is generally supplied with additional metal called filler. Filler material depends upon the metals to be welded.

In oxy-fuel cutting, a torch is used to heat metal to its kindling temperature. A stream of oxygen is then trained on the metal, burning it into a metal oxide that flows out of the kerf as slag.

Torches that do not mix fuel with oxygen (combining, instead, atmospheric air) are not considered oxy-fuel torches and can typically be identified by a single tank (oxy-fuel cutting requires two isolated supplies, fuel and oxygen). Most metals cannot be melted with a single-tank torch. Consequently, single-tank torches are typically suitable for soldering and brazing but not for welding.

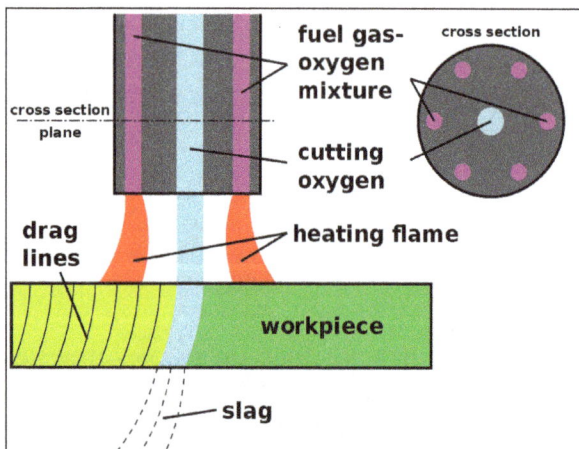

Principle of burn cutting.

Torch-cut pipe.

A cutting torch is used to cut a steel pipe.

Uses

Oxy-fuel torches are or have been used for:

- Heating metal: in automotive and other industries for the purposes of loosening seized fasteners.

- Neutral flame is used for joining and cutting of all ferrous and non ferrous metals except brass.

- Depositing metal to build up a surface, as in hardfacing.

- Also, oxy-hydrogen flames are used:

  - In stone working for "flaming" where the stone is heated and a top layer crackles and breaks. A steel circular brush is attached to an angle grinder and used to remove the first layer leaving behind a bumpy surface similar to hammered bronze.

  - In the glass industry for "fire polishing".

  - In jewelry production for "water welding" using a water torch (an oxyhydrogen torch whose gas supply is generated immediately by electrolysis of water).

  - In automotive repair, removing a seized bolt.

  - Formerly, to heat lumps of quicklime to obtain a bright white light called limelight, in theatres or optical ("magic") lanterns.

  - formerly, in platinum works, as platinum is fusible only in the oxyhydrogen flame and in an electric furnace.

In short, oxy-fuel equipment is quite versatile, not only because it is preferred for some sorts of iron or steel welding but also because it lends itself to brazing, braze-welding, metal heating (for annealing or tempering, bending or forming), rust or scale removal, the loosening of corroded nuts and bolts, and is a ubiquitous means of cutting ferrous metals.

Apparatus

The apparatus used in gas welding consists basically of an oxygen source and a fuel gas source (usually contained in cylinders), two pressure regulators and two flexible hoses (one for each cylinder), and a torch. This sort of torch can also be used for soldering and brazing. The cylinders are often carried in a special wheeled trolley.

There have been examples of oxyhydrogen cutting sets with small (scuba-sized) gas cylinders worn on the user's back in a backpack harness, for rescue work and similar.

There are also examples of pressurized liquid fuel cutting torches, usually using gasoline. These are used for their increased portability.

Regulator

The regulator ensures that pressure of the gas from the tanks matches the required pressure in the hose. The flow rate is then adjusted by the operator using needle valves on the torch. Accurate flow control with a needle valve relies on a constant inlet pressure.

Most regulators have two stages. The first stage is a fixed-pressure regulator, which releases gas from the cylinder at a constant intermediate pressure, despite the pressure in the cylinder falling as the gas in it is consumed. This is similar to the first stage of a scuba-diving regulator. The adjustable second stage of the regulator controls the pressure reduction from the intermediate pressure to the low outlet pressure. The regulator has two pressure gauges, one indicating cylinder pressure, the other indicating hose pressure. The adjustment knob of the regulator is sometimes roughly calibrated for pressure, but an accurate setting requires observation of the gauge.

Some simpler or cheaper oxygen-fuel regulators have only a single-stage regulator, or only a single gauge. A single-stage regulator will tend to allow a reduction in outlet pressure as the cylinder is emptied, requiring manual readjustment. For low-volume users, this is an acceptable simplification. Welding regulators, unlike simpler LPG heating regulators, retain their outlet (hose) pressure gauge and do not rely on the calibration of the adjustment knob. The cheaper single-stage regulators may sometimes omit the cylinder contents gauge, or replace the accurate dial gauge with a cheaper and less precise "rising button" gauge.

Gas Hoses

The hoses are designed for use in welding and cutting metal. A double-hose or twinned design can be used, meaning that the oxygen and fuel hoses are joined together. If separate hoses are used, they should be clipped together at intervals approximately 3 feet (1 m) apart, although that is not recommended for cutting applications, because beads of molten metal given off by the process can become lodged between the hoses where they are held together, and burn through, releasing the pressurised gas inside, which in the case of fuel gas usually ignites.

The hoses are color-coded for visual identification. The color of the hoses varies between countries. In the United States, the oxygen hose is green and the fuel hose is red. In the UK and other countries, the oxygen hose is blue (black hoses may still be found on old equipment), and the acetylene (fuel) hose is red. If liquefied petroleum gas (LPG) fuel, such as propane, is used, the fuel hose should be orange, indicating that it is compatible with LPG. LPG will damage an incompatible hose, including most acetylene hoses.

The threaded connectors on the hoses are handed to avoid accidental mis-connection: the thread on the oxygen hose is right-handed (as normal), while the fuel gas hose has a left-handed thread. The left-handed threads also have an identifying groove cut into their nuts.

Gas-tight connections between the flexible hoses and rigid fittings are made by using crimped hose clips or ferrules, often referred to as 'O' clips, over barbed spigots. The use of worm-drive hose clips or Jubilee clips is specifically forbidden in the UK and other countries.

Non-Return Valve

Acetylene is not just flammable; in certain conditions it is explosive. Although it has an upper flammability limit in air of 81%, acetylene's explosive decomposition behaviour makes this irrelevant. If a detonation wave enters the acetylene tank, the tank will be blown apart by the decomposition. Ordinary check valves that normally prevent back flow cannot stop a detonation wave because they are not capable of closing before the wave passes around the gate. For that reason a flashback arrestor is needed. It is designed to operate before the detonation wave makes it from the hose side to the supply side.

Between the regulator and hose, and ideally between hose and torch on both oxygen and fuel lines, a flashback arrestor and/or non-return valve (check valve) should be installed to prevent flame or oxygen-fuel mixture being pushed back into either cylinder and damaging the equipment or causing a cylinder to explode.

European practice is to fit flashback arrestors at the regulator and check valves at the torch. US practice is to fit both at the regulator.

The flashback arrestor prevents shock waves from downstream coming back up the hoses and entering the cylinder, possibly rupturing it, as there are quantities of fuel/oxygen mixtures inside parts of the equipment (specifically within the mixer and blowpipe/nozzle) that may explode if the equipment is incorrectly shut down, and acetylene decomposes at excessive pressures or temperatures. In case the pressure wave has created a leak downstream of the flashback arrestor, it will remain switched off until someone resets it.

Check Valve

A check valve lets gas flow in one direction only. It is usually a chamber containing a ball that is pressed against one end by a spring. Gas flow one way pushes the ball out of the way, and a lack of flow or a reverse flow allows the spring to push the ball into the inlet, blocking it. Not to be confused with a flashback arrestor, a check valve is not designed to block a shock wave. The shock wave could occur while the ball is so far from the inlet that the wave will get past the ball before it can reach its off position.

Torch

The torch is the tool that the welder holds and manipulates to make the weld. It has a connection and valve for the fuel gas and a connection and valve for the oxygen, a handle for the welder to grasp, and a mixing chamber (set at an angle) where the fuel gas and oxygen mix, with a tip where the flame forms. Two basic types of torches are positive pressure type and low pressure or injector type.

Welding Torch

A welding torch head is used to weld metals. It can be identified by having only one or two pipes running to the nozzle, no oxygen-blast trigger, and two valve knobs at the bottom of the handle letting the operator adjust the oxygen and fuel flow respectively.

The top torch is a welding torch and the bottom is a cutting torch.

Cutting Torch

A cutting torch head is used to cut materials. It is similar to a welding torch, but can be identified by the oxygen blast trigger or lever.

When cutting, the metal is first heated by the flame until it is cherry red. Once this temperature is attained, oxygen is supplied to the heated parts by pressing the oxygen-blast trigger. This oxygen reacts with the metal, forming an oxide and producing heat. It is the heat that continues the cutting process. The cutting torch only heats the metal to start the process; further heat is provided by the burning metal.

The melting point of the iron oxide is around half that of the metal being cut. As the metal burns, it immediately turns to liquid iron oxide and flows away from the cutting zone. However, some of the iron oxide remains on the workpiece, forming a hard "slag" which can be removed by gentle tapping and/or grinding.

Rose Bud Torch

A rose bud torch is used to heat metals for bending, straightening, etc. where a large area needs to be heated. It is so-called because the flame at the end looks like a rose bud. A welding torch can also be used to heat small areas such as rusted nuts and bolts.

Injector Torch

A typical oxy-fuel torch, called an equal-pressure torch, merely mixes the two gases. In an injector torch, high-pressure oxygen comes out of a small nozzle inside the torch head which drags the fuel gas along with it, using the venturi effect.

Fuels

Oxy-fuel processes may use a variety of fuel gases, the most common being acetylene. Other gases that may be used are propylene, liquified petroleum gas (LPG), propane, natural gas, hydrogen, and MAPP gas. Many brands use different kinds of gases in their mixes.

Acetylene

Acetylene is the primary fuel for oxy-fuel welding and is the fuel of choice for repair work and general cutting and welding. Acetylene gas is shipped in special cylinders designed to keep the gas dissolved. The cylinders are packed with porous materials (e.g. kapok fibre, diatomaceous earth, or (formerly) asbestos), then filled to around 50% capacity with acetone, as acetylene is soluble in acetone. This method is necessary because above 207 kPa (30 lbf/in²) (absolute pressure) acetylene is unstable and may explode.

There is about 1700 kPa (250 psi) pressure in the tank when full. Acetylene when combined with oxygen burns at 3200 °C to 3500 °C (5800 °F to 6300 °F), highest among commonly used gaseous fuels. As a fuel acetylene's primary disadvantage, in comparison to other fuels, is high cost.

As acetylene is unstable at a pressure roughly equivalent to 33 feet/10 meters underwater, water-submerged cutting and welding is reserved for hydrogen rather than acetylene.

Acetylene generator as used in Bali by reaction of calcium carbide with water. This is used where acetylene cylinders are not available. The term 'Las Karbit' means acetylene (carbide) welding in Indonesian.

Gasoline

Oxy-gasoline, also known as oxy-petrol, torches have been found to perform very well, especially where bottled gas fuel is not available or difficult to transport to the worksite. Tests showed that an oxy-gasoline torch can cut steel plate up to 0.5 in (13 mm) thick at the same rate as oxy-acetylene. In plate thicknesses greater than 0.5 in (13 mm) the cutting rate was better than that of oxy-acetylene; at 4.5 in (110 mm) it was three times faster.

The gasoline is fed either from a pressurised tank (whose pressure can be hand-pumped or fed from a gas cylinder).OR from a non pressurised tank with the fuel being drawn into the torch by venturi action by the pressurised oxygen flow. Another low cost approach commonly used by jewelry makers in Asia is using air bubbled through a gasoline container by a foot-operated air pump, and burning the fuel-air mixture in a specialized welding torch.

Hydrogen

Hydrogen has a clean flame and is good for use on aluminium. It can be used at a higher pressure than acetylene and is therefore useful for underwater welding and cutting. It is a good type of flame to use when heating large amounts of material. The flame temperature is high, about 2,000 °C for hydrogen gas in air at atmospheric pressure, and up to 2800 °C when pre-mixed in a 2:1 ratio with pure oxygen (oxyhydrogen). Hydrogen is not used for welding steels and other ferrous materials, because it causes hydrogen embrittlement.

For some oxyhydrogen torches the oxygen and hydrogen are produced by electrolysis of water in an apparatus which is connected directly to the torch. Types of this sort of torch:

- The oxygen and the hydrogen are led off the electrolysis cell separately and are fed into the two gas connections of an ordinary oxy-gas torch. This happens in the water torch, which is sometimes used in small torches used in making jewelry and electronics.

- The mixed oxygen and hydrogen are drawn from the electrolysis cell and are led into a special torch designed to prevent flashback.

MPS and MAPP Gas

Methylacetylene-propadiene (MAPP) gas and LPG gas are similar fuels, because LPG gas is liquefied petroleum gas mixed with MPS. It has the storage and shipping characteristics of LPG and has a heat value a little lower than that of acetylene. Because it can be shipped in small containers for sale at retail stores, it is used by hobbyists and large industrial companies and shipyards because it does not polymerize at high pressures — above 15 psi or so (as acetylene does) and is therefore much less dangerous than acetylene. Further, more of it can be stored in a single place at one time, as the increased compressibility allows for more gas to be put into a tank. MAPP gas can be used at much higher pressures than acetylene, sometimes up to 40 or 50 psi in high-volume oxy-fuel cutting torches which can cut up to 12-inch-thick (300 mm) steel. Other welding gases that develop comparable temperatures need special procedures for safe shipping and handling. MPS and MAPP are recommended for cutting applications in particular, rather than welding applications.

On 31 April 2008 the Petromont Varennes plant closed its methylacetylene/propadiene crackers. As it was the only North American plant making MAPP gas, many substitutes were introduced by companies that had repackaged the Dow and Varennes products - most of these substitutes are propylene.

Propylene and Fuel Gas

Propylene is used in production welding and cutting. It cuts similarly to propane. When propylene is used, the torch rarely needs tip cleaning. There is often a substantial advantage to cutting with an injector torch rather than an equal-pressure torch when using propylene. Quite a few North American suppliers have begun selling propylene under proprietary trademarks such as FG2 and Fuel-Max.

Butane, Propane and Butane/Propane Mixes

Butane, like propane, is a saturated hydrocarbon. Butane and propane do not react with each other and are regularly mixed. Butane boils at 0.6 °C. Propane is more volatile, with a boiling point of -42 °C. Vaporization is rapid at temperatures above the boiling points. The calorific (heat) values of the two are almost equal. Both are thus mixed to attain the vapor pressure that is required by the end user and depending on the ambient conditions. If the ambient temperature is very low, propane is preferred to achieve higher vapor pressure at the given temperature.

Propane does not burn as hot as acetylene in its inner cone, and so it is rarely used for welding. Propane, however, has a very high number of BTUs per cubic foot in its outer cone, and so with the right torch (injector style) can make a faster and cleaner cut than acetylene, and is much more useful for heating and bending than acetylene. The maximum neutral flame temperature of propane in oxygen is 2,822 °C (5,112 °F). Propane is cheaper than acetylene and easier to transport.

Role of Oxygen

Oxygen is not the fuel. It is what chemically combines with the fuel to produce the heat for welding. This is called 'oxidation', but the more specific and more commonly used term in this context is 'combustion'. In the case of hydrogen, the product of combustion is simply water. For the other hydrocarbon fuels, water and carbon dioxide are produced. The heat is released because the molecules of the products of combustion have a lower energy state than the molecules of the fuel and oxygen. In oxy-fuel cutting, oxidation of the metal being cut (typically iron) produces nearly all of the heat required to "burn" through the workpiece.

Oxygen is usually produced elsewhere by distillation of liquefied air and shipped to the welding site in high-pressure vessels (commonly called "tanks" or "cylinders") at a pressure of about 21,000 kPa (3,000 lbf/in^2 = 200 atmospheres). It is also shipped as a liquid in Dewar type vessels (like a large Thermos jar) to places that use large amounts of oxygen.

It is also possible to separate oxygen from air by passing the air, under pressure, through a zeolite sieve that selectively adsorbs the nitrogen and lets the oxygen (and argon) pass. This gives a purity of oxygen of about 93%. This method works well for brazing, but higher-purity oxygen is necessary to produce a clean, slag-free kerf when cutting.

Types of Flame

The welder can adjust the oxy-acetylene flame to be carbonizing (aka reducing), neutral, or oxidizing. Adjustment is made by adding more or less oxygen to the acetylene flame. The neutral flame is the flame most generally used when welding or cutting. The welder uses the neutral flame as the starting point for all other flame adjustments because it is so easily defined. This flame is attained when welders, as they slowly open the oxygen valve on the torch body, first see only two flame zones. At that point, the acetylene is being completely burned in the welding oxygen and surrounding air. The flame is chemically neutral. The two parts of this flame are the light blue inner cone and the darker blue to colorless outer cone. The inner cone is where the acetylene and the oxygen combine. The tip of this inner cone is the hottest part of the flame. It is approximately 6,000 °F (3,300 °C) and provides enough heat to easily melt steel. In the inner cone the acetylene breaks down and partly burns to hydrogen and carbon monoxide, which in the outer cone combine with more oxygen from the surrounding air and burn.

An excess of acetylene creates a carbonizing flame. This flame is characterized by three flame zones; the hot inner cone, a white-hot "acetylene feather", and the blue-colored outer cone. This is the type of flame observed when oxygen is first added to the burning acetylene. The feather is adjusted and made ever smaller by adding increasing amounts of oxygen to the flame. A welding feather is measured as 2X or 3X, with X being the length of the inner flame cone. The unburned carbon insulates the flame and drops the temperature to approximately 5,000 °F (2,800 °C). The reducing flame is typically used for hard facing operations or backhand pipe welding techniques. The feather is caused by incomplete combustion of the acetylene to cause an excess of carbon in the flame. Some of this carbon is dissolved by the molten metal to carbonize it. The carbonizing flame will tend to remove the oxygen from iron oxides which may be present, a fact which has caused the flame to be known as a "reducing flame".

The oxidizing flame is the third possible flame adjustment. It occurs when the ratio of oxygen to acetylene required for a neutral flame has been changed to give an excess of oxygen. This flame type is observed when welders add more oxygen to the neutral flame. This flame is hotter than the other two flames because the combustible gases will not have to search so far to find the necessary amount of oxygen, nor heat up as much thermally inert carbon. It is called an oxidizing flame because of its effect on metal. This flame adjustment is generally not preferred. The oxidizing flame creates undesirable oxides to the structural and mechanical detriment of most metals. In an oxidizing flame, the inner cone acquires a purplish tinge and gets pinched and smaller at the tip, and the sound of the flame gets harsh. A slightly oxidizing flame is used in braze-welding and bronze-surfacing while a more strongly oxidizing flame is used in fusion welding certain brasses and bronzes.

The size of the flame can be adjusted to a limited extent by the valves on the torch and by the regulator settings, but in the main it depends on the size of the orifice in the tip. In fact, the tip should be chosen first according to the job at hand, and then the regulators set accordingly.

Welding

The flame is applied to the base metal and held until a small puddle of molten metal is formed. The puddle is moved along the path where the weld bead is desired. Usually, more metal is added to the puddle as it is moved along by dipping metal from a welding rod or filler rod into the molten metal puddle. The metal puddle will travel towards where the metal is the hottest. This is accomplished through torch manipulation by the welder.

The amount of heat applied to the metal is a function of the welding tip size, the speed of travel, and the welding position. The flame size is determined by the welding tip size. The proper tip size is determined by the metal thickness and the joint design.

Welding gas pressures using oxy-acetylene are set in accordance with the manufacturer's recommendations. The welder will modify the speed of welding travel to maintain a uniform bead width. Uniformity is a quality attribute indicating good workmanship. Trained welders are taught to keep the bead the same size at the beginning of the weld as at the end. If the bead gets too wide, the welder increases the speed of welding travel. If the bead gets too narrow or if the weld puddle is lost, the welder slows down the speed of travel. Welding in the vertical or overhead positions is typically slower than welding in the flat or horizontal positions.

The welder must add the filler rod to the molten puddle. The welder must also keep the filler metal in the hot outer flame zone when not adding it to the puddle to protect filler metal from oxidation.

Do not let the welding flame burn off the filler metal. The metal will not wet into the base metal and will look like a series of cold dots on the base metal. There is very little strength in a cold weld. When the filler metal is properly added to the molten puddle, the resulting weld will be stronger than the original base metal.

Welding lead or 'lead burning' was much more common in the 19th century to make some pipe connections and tanks. Great skill is required but can be quickly learned. In building construction today some lead flashing is welded but soldered copper flashing is much more common in America. In the automotive body collision industry before the 1980s, oxyacetylene gas torch welding was seldom used to weld sheetmetal, since warpage was a byproduct besides the excess heat. Automotive body repair methods at the time were crude and yielded improprieties until MIG welding became the industry standard. Since the 1970s, when high strength steel became the standard for automotive manufacturing, electric welding became the preferred method. After the 1980s, the oxyacetylene torch fell out of use for sheetmetal welding in the industrialized world.

Cutting

For cutting, the setup is a little different. A cutting torch has a 60- or 90-degree angled head with orifices placed around a central jet. The outer jets are for preheat flames of oxygen and acetylene. The central jet carries only oxygen for cutting. The use of several preheating flames rather than a single flame makes it possible to change the direction of the cut as desired without changing the position of the nozzle or the angle which the torch makes with the direction of the cut, as well as giving a better preheat balance. Manufacturers have developed custom tips for Mapp, propane, and propylene gases to optimize the flames from these alternate fuel gases.

The flame is not intended to melt the metal, but to bring it to its ignition temperature.

The torch's trigger blows extra oxygen at higher pressures down the torch's third tube out of the central jet into the workpiece, causing the metal to burn and blowing the resulting molten oxide through to the other side. The ideal kerf is a narrow gap with a sharp edge on either side of the workpiece; overheating the workpiece and thus melting through it causes a rounded edge.

Oxygen Rich Butane Torch Flame.

Fuel Rich Butane Torch Flame.

Cutting is initiated by heating the edge or leading face (as in cutting shapes such as round rod) of the steel to the ignition temperature (approximately bright cherry red heat) using the pre-heat jets only, then using the separate cutting oxygen valve to release the oxygen from the central jet. The oxygen chemically combines with the iron in the ferrous material to oxidize the iron quickly into molten iron oxide, producing the cut. Initiating a cut in the middle of a workpiece is known as piercing.

Cutting a rail just before renewing the rails and the ballast.

It is worth noting several things at this point:

- The oxygen flowrate is critical; too little will make a slow ragged cut, while too much will waste oxygen and produce a wide concave cut. Oxygen lances and other custom made torches do not have a separate pressure control for the cutting oxygen, so the cutting oxygen pressure must be controlled using the oxygen regulator. The oxygen cutting pressure should match the cutting tip oxygen orifice. Consult the tip manufacturer's equipment data for the proper cutting oxygen pressures for the specific cutting tip.

- The oxidation of iron by this method is highly exothermic. Once it has started, steel can be cut at a surprising rate, far faster than if it were merely melted through. At this point, the pre-heat jets are there purely for assistance. The rise in temperature will be obvious by the intense glare from the ejected material, even through proper goggles. (A thermic lance is a tool that also uses rapid oxidation of iron to cut through almost any material.)

- Since the melted metal flows out of the workpiece, there must be room on the opposite side of the workpiece for the spray to exit. When possible, pieces of metal are cut on a grate that lets the melted metal fall freely to the ground. The same equipment can be used for oxyacetylene blowtorches and welding torches, by exchanging the part of the torch in front of the torch valves.

For a basic oxy-acetylene rig, the cutting speed in light steel section will usually be nearly twice as fast as a petrol-driven cut-off grinder. The advantages when cutting large sections are obvious: an oxy-fuel torch is light, small and quiet and needs very little effort to use, whereas a cut-off grinder is heavy and noisy and needs considerable operator exertion and may vibrate severely, leading to stiff hands and possible long-term vibration white finger. Oxy-acetylene torches can easily cut through ferrous materials in excess of 200 mm (8 inches). Oxygen lances are used in scrapping operations and cut sections thicker than 200 mm (8 inches). Cut-off grinders are useless for these kinds of application.

Robotic oxy-fuel cutters sometimes use a high-speed divergent nozzle. This uses an oxygen jet that opens slightly along its passage. This allows the compressed oxygen to expand as it leaves, forming a high-velocity jet that spreads less than a parallel-bore nozzle, allowing a cleaner cut. These are not used for cutting by hand since they need very accurate positioning above the work. Their ability to produce almost any shape from large steel plates gives them a secure future in shipbuilding and in many other industries.

Oxy-propane torches are usually used for cutting up scrap to save money, as LPG is far cheaper joule for joule than acetylene, although propane does not produce acetylene's very neat cut profile. Propane also finds a place in production, for cutting very large sections.

Oxy-acetylene can cut only low- to medium-carbon steels and wrought iron. High-carbon steels are difficult to cut because the melting point of the slag is closer to the melting point of the parent metal, so that the slag from the cutting action does not eject as sparks but rather mixes with the clean melt near the cut. This keeps the oxygen from reaching the clean metal and burning it. In the case of cast iron, graphite between the grains and the shape of the grains themselves interfere with the cutting action of the torch. Stainless steels cannot be cut either because the material does not burn readily.

Health Hazards of Oxy-Acetylene Cutting and Welding

Oxy-acetylene cutting – also known as gas cutting or oxy-fuel cutting – is often considered less hazardous than traditional welding. Accordingly, the health hazards are often ignored. However, most or all of the hazards of welding may still exist. In gas cutting, a torch is used to heat metal to its auto-ignition temperature and a stream of oxygen is then trained on the metal, turning it into a metal oxide that flows out as slag. Acetylene is the most common gas used for fueling cutting torches. When mixed with pure oxygen in a cutting torch assembly, an acetylene flame can theoretically reach over 5700°F.

In general industry, oxygen gas cutting may be part of the work process where hazards are determined and controls are in place. But more often, it is done as maintenance operations or non-routine work. In construction, it is often done as demolition work. In these cases, ventilation may not be available, the intensity and duration of the work may be vary, and the hazards undetermined.

For example, on a construction job where employees torch cut painted metal, personal sampling levels may far exceed occupational exposure standards for lead.

Health hazards of gas cutting are due to the radiation and toxic fumes or gases emitted during the process. The fume created by gas cutting is a mixture of very fine particles and gases. Many substances, such as chromium, nickel, arsenic, asbestos, manganese, silica, beryllium, cadmium, nitrogen oxides, phosgene, acrolein, fluorine compounds, carbon monoxide, cobalt, copper, lead, ozone, selenium, and zinc, may be present and can be extremely toxic.

Resultant health problems from gas cutting may include:

- Eye injuries, such as discomfort and burns from the intense light and heat emitted from the operation, and cataracts caused by radiation from molten metal, leading to inability to see things clearly, or corneal ulcer and conjunctivitis from foreign particles e.g. slag and cutting sparks.

- Skin irritation and reddening due to over exposure to radiation.

- Illness due to inhalation of fumes or gases formed during the process, such as metal fume fever from freshly formed metal oxide, illness from toxic fumes of metals such as lead, cadmium, beryllium, bronchial and pulmonary irritation from toxic gases such as oxides of nitrogen, fluorides; burns from the flame, hot slag or hot surfaces of the work.

- Asphyxiation due work in confined space which may deplete oxygen levels.

- Heat-stroke from prolonged operation with the flame, especially in confined space.

Safety Precautions in Oxy Acetylene Welding

Safe Storage

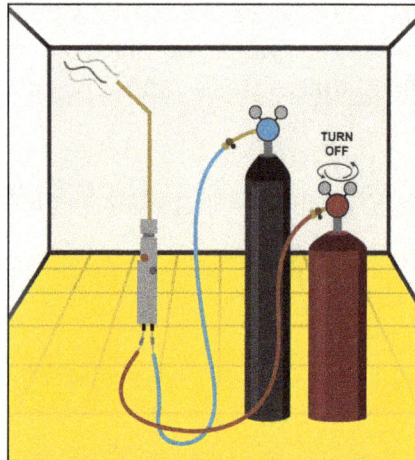

Gases are normally supplied under high pressure in steel cylinders; in the UK, the colour coding for the cylinders is in the process of being harmonised across Europe. For acetylene the shoulder of the cylinder is maroon and for oxygen the shoulder is white, although black oxygen cylinders will remain in circulation for some time. The cylinder should also have a label marked with the type of gas. To prevent the interchange of fittings between cylinders containing combustible and non-combustible gases, oxygen cylinders have a right-hand and acetylene have a left-hand thread. All cylinders are opened by turning the key or knob anticlockwise and closed by turning them clockwise.

Oxygen will cause a fire to burn more fiercely and a mixture of oxygen and a fuel gas can cause an explosion. It is, therefore, essential that the oxygen cylinders are separated from the fuel gas cylinders and stored in an area free from combustible material.

Safe Practice and Accident Avoidance

- Store the cylinders in a well-ventilated area, preferably in the open air.

- The storage area should be well away from sources of heat, sparks and fire risk.

- Cylinders should be stored upright and well secured.

- Oxygen cylinders should be stored at least 3m from fuel gas cylinders or separated by a 30 minute fire resisting barrier.

- The store area should be designated 'No Smoking'.

Handling Compressed Gases

Cylinders are fitted with regulators to reduce the gas pressure in the cylinder to the working pressure of the torch. The regulator has two gauges, a high pressure gauge for the gas in the cylinder and a low pressure gauge for the gas being fed to the torch. The gas flow rate is controlled by a pressure adjusting screw which sets the outlet gas pressure. The BCGA Code of Practice CP7 rec-ommends the gauges are checked annually and replaced every 5 years.

Factors to be considered are that the gas system is suitable for the pressure rating and the hoses are connected without any leaks. Valve threads should be cleaned before screwing in the regulator. The valve of an acetylene cylinders can be opened slightly to blow out the threads but the threads in oxygen cylinders are best cleaned using clean compressed air (the threads on hydrogen cylinders must always be blown out using compressed air).

As oxygen can react violently with oils and grease, lubricating oils or sealant for the threads must not be used.

- Cylinders are very heavy and must be securely fastened at all times.

- Cylinder valves or valve guards should never be loosened.

- Check the regulator is rated for the pressure in the cylinder.

- When attaching the regulator to the cylinder the joints must be clean and sealant must not be used.

- Before attaching a regulator, the pressure adjustment screw must be screwed out to prevent unregulated flow of gas into the system when the cylinder valve is opened.

Using Compressed Gases

Gases are mixed in the hand-held torch or blowpipe in the correct proportions. Hoses between regulator and torch should be colour coded; in the UK, red for acetylene and blue for oxygen. Hoses should be kept as short as possible and users should check periodically that they are not near hot or sharp objects which could damage the hose wall. Acetylene cylinders must always be used upright.

When connecting the system, and at least at the start of each shift, hoses and torch must be purged to remove any inflammable gas mixtures. It is essential the oxygen stream does not come into contact with oil which can ignite spontaneously. Purging should also not be carried out in confined spaces.

The torch should be lit with a friction lighter or stationary pilot flame to avoid burning the hands; matches should not be used and the flame should not be reignited from hot metal, especially when working in a confined space.

The cylinders should not become heated, for example by allowing the torch flame to heat locally the cylinder wall. Similarly, arc welding too close to the cylinder could result in an arc forming between the cylinder and workpiece/electrode.

Although very little UV is emitted, the welder must wear tinted goggles. The grade of filter is determined by the intensity of the flame which depends on the thickness of metal being welded; recommendations for filters according to the acetylene flow rate are given in the table.

Grade of filter recommended according to the acetylene flow rate:

Work	flow rate of acetylene in l/hr			
	up to 70	70 - 200	200 - 800	over 800
Welding and braze welding of heavy metals e.g. steels, copper and their alloys	4	5	6	7
Welding with emittive fluxes (notably light alloys)	4a	5a	6a	7a

Safe practice and accident avoidance

- When cleaning the cylinder threads, connecting the regulator and purging the hoses, protect face and eyes by wearing the appropriate head shield.

- Use a suitable welding shield equipped with the appropriate ocular protection filter.

- Wear non-combustible clothing.

- Ensure the cylinder is not heated by the flame or by stray arcs from adjacent electrical equipment.

Leak detection

Joints and hoses should be checked for leaks before any welding is attempted. Whilst acetylene may be detected by its distinctive smell (usually at levels of less than 2%) oxygen is odourless.

Leak detection is best carried out applying a weak (typically 0.5%) solution of a detergent in water or a leak detecting solution from one of the gas supply companies. It is applied to the joints using a brush and the escaping gas will form bubbles. On curing the leak, the area should be cleaned to remove the residue from the leak detecting solution. Leaks in hoses may be repaired but approved replacement hose and couplings must be used in accordance with BSEN 560:1995 and BSEN 1256:1996.

Backfire and Flashback

A backfire (a single cracking or 'popping' sound) is when the flame has ignited the gases inside the nozzle and extinguished itself. This may happen when the torch is held too near the workpiece.

A flashback (a shrill hissing sound) when the flame is burning inside the torch, is more severe. The flame may pass back through the torch mixing chamber to the hose. The most likely cause is incorrect gas pressures giving too low a gas velocity. Alternatively, a situation may be created by a higher pressure gas (acetylene) feeding up a lower pressure gas (oxygen) stream. This could occur if the oxygen cylinder is almost empty but other potential causes would be hose leaks, loose connections, or failure to adequately purge the hoses.

Non-return valves fitted to the hoses will detect and stop reverse gas flow preventing an inflammable oxygen and acetylene mixture from forming in the hose. The flashback arrestor is an automatic flame trap device designed not only to quench the flame but also to prevent the flame from reaching the regulator.

Backfire or Flashback Procedure

After an unsustained backfire in which the flame is extinguished:

- Close the blowpipe control valves (fuel gas first).
- Check the nozzle is tight.
- Check the pressures on regulators.
- Re-light the torch using the recommended procedure.

If the flame continues to burn:

- Close the oxygen valve at the torch (to prevent internal burning).
- Close the acetylene valve at the torch.
- Close cylinder valves or gas supply point isolation valves for both oxygen and acetylene.
- Close outlets of adjustable pressure regulators by winding out the pressure-adjusting screws.
- Open both torch valves to vent the pressure in the equipment.
- Close torch valves.
- Check nozzle tightness and pressures on regulators.
- Re-light the torch using the recommended procedure.

If a flashback occurs in the hose and equipment, or fire in the hose, regulator connections or gas supply outlet points:

- Isolate oxygen and fuel gas supplies at the cylinder valves or gas supply outlet points (only if this can be done safely).
- If no risk of personal injury, control fire using first aid fire-fighting equipment.
- If the fire cannot be put out at once, call emergency fire services.
- After the equipment has cooled, examine the equipment and replace defective components.

When a backfire has been investigated and the fault rectified, the torch may be re-lit. After a flashback, because the flame has extended to the regulator it is essential not only to examine the torch, but the hoses and components must be checked and, if necessary, replaced. The flashback arrestor should also be checked according to manufacturer's instructions and, with some designs, it may be necessary to replace it.

Oxy Hydrogen Welding

In oxy-hydrogen welding, hydrogen combines with oxygen to generate steam and attains a flame temperature of around 2800°C. But the weld pool is not protected from the atmosphere when the oxygen for combustion is completely provided by the torch itself. So, Oxygen is an amount slightly less than that required for complete combustion is provided by the torch, whereas atmospheric oxygen accounts for the burning of the remaining hydrogen. This gives rise to a protective preheating flame that surrounds the main flame. But this reduces the flame temperature to some extent. Because of the lower flame temperature, oxy-hydrogen welding is generally slow process. It is normally used to weld thin sheets of steels and alloys with low melting temperatures.

Key Features

- Operation is convenient & safe: Oxy hydrogen generator produce oxygen and hydrogen gas that you required and also no gas cylinder is required. There is no risk of explosion.

- Environmental friendly: In this process fuel comes from water and there is water vapour after finishing this process so this process is environmental friendly.

- Welding features: Welding work is fast, precision, smooth and beautiful welding spot. Oxy-hydrogen flame is concentrated up to 2800 °C so it can heat the welding spot to melting point very quickly.

- Energy saving and Low cost: This process is done by very low electricity and pure water. The cost of electricity and water is reduced more than 40% compared with LPG and other welding process.

References

- Gas-welding-principle-working-equipment: mech4study.com, Retrieved 12 April, 2019

- Carlisle, Rodney (2004). Scientific American Inventions and Discoveries, p.365. John Wright & Songs, Inc., New Jersey. ISBN 0-471-24410-4

- Welding-flames, OLDSITE: weldguru.com, Retrieved 13 May, 2019

- "Portable Oxy-Fuel Gas Equipment" (PDF). Virginia Polytechnic Institute and State University. Retrieved 2016-02-02

- Health-hazards-of-oxy-acetylene-cutting-welding-june-2015, newsletter: occusafeinc.com, Retrieved 14 June, 2019

- Health-safety-and-accident-prevention-oxyacetylene-welding-cutting-and-heating-027, job-knowledge, technical-knowledge: twi-global.com, Retrieved 15 July, 2019

3
Arc Welding

Arc welding refers to the welding process that uses electricity for creating heat to melt and join metals. Shielded metal arc welding, flux cored arc welding, gas tungsten arc welding, plasma arc welding, submerged arc welding, etc. are a few of its types. This chapter closely examines these different types of arc welding to provide an extensive understanding of the subject.

Arc welding is the fusion of two pieces of metal by an electric arc between the pieces being joined the work pieces – and an electrode that is guided along the joint between the pieces. The electrode is either a rod that simply carries current between the tip and the work, or a rod or wire that melts and supplies filler metal to the joint.

The basic arc welding circuit is an alternating current (AC) or direct current (DC) power source connected by a "work" cable to the work piece and by a "hot" cable to an electrode. When the electrode is positioned close to the work piece, an arc is created across the gap between the metal and the hot cable electrode. An ionized column of gas develops to complete the circuit.

Basic Welding Circuit

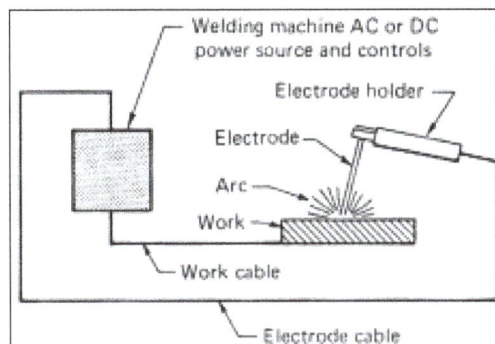

Basic Welding Circuit.

The arc produces a temperature of about 3600 °C at the tip and melts part of the metal being welded and part of the electrode. This produces a pool of molten metal that cools and solidifies behind the electrode as it is moved along the joint.

There are two types of electrodes. Consumable electrode tips melt, and molten metal droplets detach and mix into the weld pool. Non-consumable electrodes do not melt. Instead, filler metal is melted into the joint from a separate rod or wire.

The strength of the weld is reduced when metals at high temperatures react with oxygen and nitrogen in the air to form oxides and nitrides. Most arc welding processes minimize contact between the molten metal and the air with a shield of gas, vapour or slag. Granular flux, for example, adds deoxidizers that create a shield to protect the molten pool, thus improving the weld.

Advances in Welding Power Source Design and Efficiency

The electricity-consuming device – the key component of the arc welding apparatus – is the power source. Electrical consumption from the approximately 110 000 to 130 000 arc welding machines in use in Canada is estimated at 100 GWh a year.

In the past, power sources used transformer-rectifier equipment with large step-down transformers that made them heavy and prone to overheating. They can be used for only one function, i.e., one type of welding. In the 1990s, advances in power switching semiconductors led to the development of inverter power sources that are multi-functional, lighter, more flexible and that provide a superior arc.

Welding power sources use electricity when welding (arc-on) and when idling. Earlier transformerrectifier equipment had energy conversion efficiencies that ranged from 40 to 60 percent and required idling power consumption of 2 to 5 kW. Modern inverter power sources have energy conversion efficiencies near 90 percent, with idling power consumption in the order of 0.1 kW.

Modern inverter power sources are gradually replacing transformer-rectifier units. They combine a quick return on investment, and, compared with transformer-rectifier units, are far more portable and easier to operate, are multi-functional rather than mono-functional, create superior arcs and combine higher-quality welds with longer arc-on time.

The Five most Common Arc Welding Processes

Process	Known as	Electrodes	Shielding	Operator Skill required	Popularity
Shielded metal arc welding	SMAW or stick	Rigid metal	Stick coatings	Low	Diminishing
Gas metal arc welding	GMAW or MIG	Solid wire	CO_2 gas	Low	Growing
Flux core arc welding	FCAW or MIG	Hollow wire	Core materials	Low	Growing
Gas tungsten arc welding	GTAW or TIG	Tungsten	Argon gas	High	Steady
Submerged arc welding	SAW	Solid wire	Argon gas	High	Steady

Power sources produce DC with the electrode either positive or negative, or AC. The choice of current and polarity depends on the process, the type of electrode, the arc atmosphere and the metal being welded.

Energy Efficiency of the Power Source

- Modern inverter power sources have high energy-conversion efficiencies and can be 50 percent more efficient than transformer-rectifier power sources.

- Modern inverter power sources for idling power requirements are 1/20th of conventional transformer-rectifier power sources.

- Modern inverter power sources have power factors that are close to 100 percent; transformerrectifier power source percentages are much lower, which reduces electricity consumption.

- Modern inverter power sources are four times lighter and much smaller than transformer-rectifier power sources. They are thus more portable and can be moved by one person instead of four, making it possible to bring the welding equipment to the job, not vice versa.

- Modern inverter power sources are multi-functional and can be used for GMAW, FCAW, SMAW and GTAW.

Operation Tips

Arc welding requires an operator and a power source. Both the operator and the equipment have roles to play in making the welding process more energy efficient.

Some Important Definitions

- Arc-on time: When the welder holds an arc between the electrode and the work piece.

- Idling time: When welding equipment is ready for use but is not generating an arc Operating factor: The ratio of arc-on time to the total time worked, often expressed as a percentage:

$$\text{Operating factor} = \frac{\text{Arc-on time}}{\text{Arc-on time} + \text{Idling time}} \times 100\%$$

- Work time: Convention is to assume total annual work time of 4000 hours (two shifts).

Power Efficiency

Welding power sources draw power when idling. Efficiency is greater when idling is reduced and the operating factor is close to 100 percent. The higher the operating factor, the more efficient the process. The following are ways to improve efficiency:

- Use the most efficient welding process: Use gas metal arc welding (GMAW) instead of shielded metal arc welding (SMAW). Typically, operating factors for SMAW fall between 10 to 30 percent; operating factors for GMAW fall between 30 to 50 percent.

- Use multi-process inverter power sources: Modern inverter power sources can be used for several welding processes and save time and effort when switching processes. For example, the Miller XTM 304 can be used for GMAW, FCAW, SMAW and GTAW.

- Automate when possible: Manage repetitive operations by applying advances in automation and computer programming.

- Reduce idling time: Cut the time spent on pre-welding tasks such as assembly, positioning, tacking and cleaning, and on follow-up operations, such as slag removal and defect repair.

- Position the work to allow down-hand welding: Experience has shown that down-hand (vertical high to low) welding is faster, easier on the operator and more error-free than other techniques.

- Train the welder: Well-trained welders work better and faster and are usually conscious of energy savings opportunities.

Power Source Performance

Certain characteristics determine the energy efficiency of power sources:

- Power factor: Power factor is the ratio of "real" electrical power made available by the welding power source for producing a welding arc (the power you can use) to the "apparent" electrical power supplied by the utility (the power you pay for). The older technology of transformer-rectifier power sources can have power factors in the order of 75 percent; modern inverter power sources have power factors close to 100 percent.

- Arc-on power and idling power: Transformer-rectifier power sources use more power in arc-on and idling modes than modern inverter power sources do with the same output.

The following table shows that the average annual electrical energy required by a typical transformerrectifier source is five to nine times the energy required by an inverter power source for the same job. In other words, the inverter source uses only 10 to 20 percent of the power needed by a transformerrectifier source.

Power Source	Process	Apparent Arc-On Power (kW)	Apparent Idling Power (kW)	Operating Factor (OF)	Annual Energy Required (kWh)
Transformer – rectifier	SMAW	10.26	4.86	10%	18 600
	(stick)	10.26	4.86	30%	25 920
Inverter	SMAW	3.91	0.12	10%	1 996
	(stick)	3.91	0.12	30%	5 028

To compare the performance of power sources use the following formula:

$$\text{Energy conversion efficiency} = \frac{\text{volt-ampere output}}{\text{volt-ampere input}}$$

The kVA input and output values for power sources at rated outputs can be found in manufacturers' equipment data sheets.

Common Electric arc Welding Processes

Shielded Metal arc Welding

Shielded Metal Arc Welding, also known as manual metal arc welding, stick welding, or electric arc welding, is the most widely used of the various arc welding processes. Welding is performed with the heat of an electric arc that is maintained between the end of a coated metal electrode and the work piece.

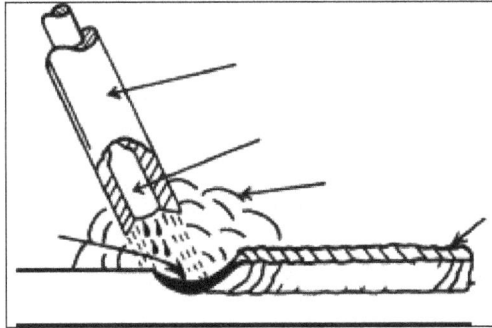

The heat produced by the arc melts the base metal, the electrode core rod, and the coating. As the molten metal droplets are transferred across the arc and into the molten weld puddle, they are shielded from the atmosphere by the gases produced from the decomposition of the flux coating. The molten slag floats to the top of the weld puddle where it protects the weld metal from the atmosphere during solidification.

Other functions of the coating are to provide arc stability and control bead shape. More information on coating functions will be covered in subsequent lessons.

Equipment & Operation: One reason for the wide acceptance of the SMAW process is the simplicity of the necessary equipment. The equipment consists of the following items.

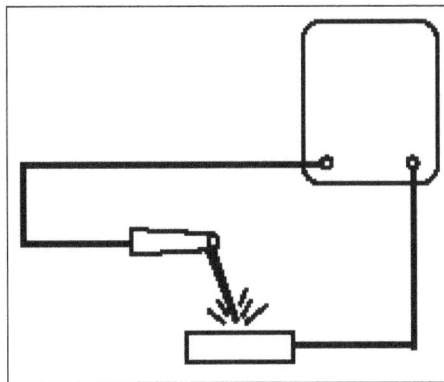

- Welding power source.

- Electrode holder.

- Ground clamp.

- Welding cables and connectors.

- Accessory equipment (chipping hammer, wire brush).

- Protective equipment (helmet, gloves, etc).

Welding Power Sources: Shielded metal arc welding may utilize either alternating current (AC) or direct current (DC), but in either case, the power source selected must be of the constant current type. This type of power source will deliver a relatively constant amperage or welding current regardless of arc length variations by the operator. The amperage determines the amount of heat at the arc and since it will remain relatively constant, the weld beads produced will be uniform in size and shape. Whether to use an AC, DC, or AC/DC power source depends on the type of welding to be done and the electrodes used. The following factors should be considered:

Electrode Selection: Using a DC power source allows the use of a greater range of electrode types. While most of the electrodes are designed to be used on AC or DC, some will work properly only on DC.

Metal Thickness: DC power sources may be used for welding both heavy sections and light gauge work. Sheet metal is more easily welded with DC because it is easier to strike and maintain the DC arc at low currents.

Distance from Work: If the distance from the work to the power source is great, AC is the best choice since the voltage drop through the cables is lower than with DC. Even though welding cables are made of copper or aluminium (both good conductors), the resistance in the cables becomes greater as the cable length increases. In other words, a voltage reading taken between the electrode and the work will be somewhat lower than a reading taken at the output terminals of the power source. This is known as voltage drop.

Welding Position: Because DC may be operated at lower welding currents, it is more suitable for overhead and vertical welding than AC. AC can successfully be used for out-of-position work if proper electrodes are selected.

Arc Blow: When welding with DC, magnetic fields are set up throughout the weldment. In weldments that have varying thickness and protrusions, this magnetic field can affect the arc by making it stray or fluctuate in direction. This condition is especially troublesome when welding in corners. AC seldom causes this problem because of the rapidly reversing magnetic field produced. Combination power sources that produce both AC and DC are available and provide the versatility necessary to select the proper welding current for the application. When using a DC power source, the question of whether to use electrode negative or positive polarity arises. Some electrodes operate on both DC straight and reverse polarity, and others on DC negative or DC positive polarity only. Direct current flows in one direction in an electrical circuit and the direction of current flow and the composition of the electrode coating will have a definite effect on the welding arc and weld bead.

The figure below shows the connections and effects of straight and reverse polarity.

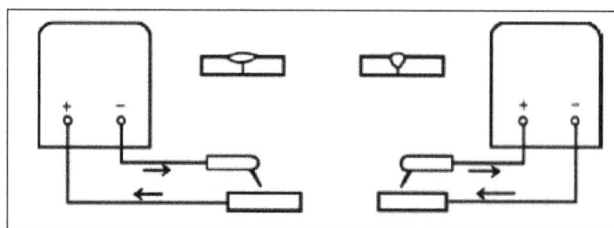

While polarity affects the penetration and burn-off rate, the electrode coating also has a strong influence on arc characteristics. Performance of individual electrodes will be discussed in succeeding lessons.

Electrode Holder: The electrode holder connects to the welding cable and conducts the welding current to the electrode. The insulated handle is used to guide the electrode over the weld joint and feed the electrode over the weld joint and feed the electrode into the weld puddle as it is consumed. Electrode holders are available in different sizes and are rated on their current carrying capacity.

Ground Clamp: The ground clamp is used to connect the ground cable to the work piece. It may be connected directly to the work or to the table or fixture upon which the work is positioned. Being a part of the welding circuit, the ground clamp must be capable of carrying the welding current without overheating due to electrical resistance.

Welding Cables: The electrode cable and the ground cable are important parts of the welding circuit. They must be very flexible and have a tough heat-resistant insulation. Connections at the electrode holder, the ground clamp, and at the power source lugs must be soldered or well crimped to assure low electrical resistance. The cross-sectional area of the cable must be sufficient size to carry the welding current with a minimum of voltage drop. Increasing the cable length necessitates increasing the cable diameter to lessen resistance and voltage drop.

Coated Electrodes: Various types of coated electrodes are used in shielded metal arc welding. Electrodes used for welding mild or carbon steels are quite different than those used for welding the low alloys and stainless steels. Details on the specific types will be covered in subsequent lessons. Gas Tungsten Arc Welding is a welding process performed using the heat of an arc established between a nonconsumable tungsten electrode and the work piece.

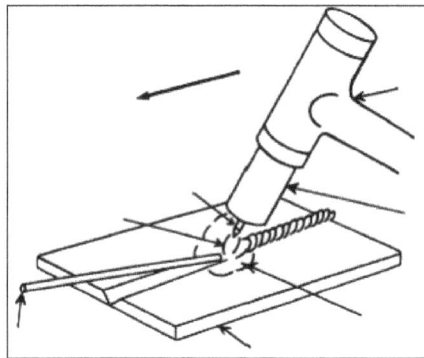

The electrode, the arc, and the area surrounding the molten weld puddle are protected from the atmosphere by an inert gas shield. The electrode is not consumed in the weld puddle as in shielded metal arc welding. If a filler metal is necessary, it is added to the leading the molten puddle. Gas tungsten arc welding produces exceptionally clean welds no slag is produced, the chance inclu-sions in the weld metal is and the finished weld requires virtually no cleaning. Argon and Helium, the primary shielding gases employed, are inert gases. Inert gases do not chemically combine with other elements and therefore, are used to exclude the reactive gases, such as oxygen and nitrogen, from forming compounds that could be detrimental to the weld metal. Gas tungsten arc welding may be used for welding almost all metals — mild steel, low alloys, stainless steel, copper and copper alloys, aluminium and aluminium alloys, nickel and nickel alloys, magnesium and magnesium alloys, titanium, and others. This process is most extensively used for welding aluminium and stainless steel alloys where weld integrity is of the utmost importance. Another use is for the root pass (initial pass) in pipe welding, which requires a weld of the highest quality. Full penetration without an excessively high inside bead is important in the root pass, and due to the ease of current control of this process, it lends itself to control of back-bead size. For high quality welds, it is usually necessary to provide an inert shielding gas inside the pipe to prevent oxidation of the inside weld bead.

Gas tungsten arc welding lends itself to both manual and automatic operation. In manual operation, the welder holds the torch in one hand and directs the arc into the weld joint. The filler metal is fed manually into the leading edge of the puddle. In automatic applications, the torch may be automatically moved over a stationary work piece or the torch may be stationary with the work moved or rotated in relation to the torch. Filler metal, if required, is also fed automatically.

Equipment and Operation - Gas tungsten arc welding may be accomplished with relatively simple equipment, or it may require some highly sophisticated components. Choice of equipment depends upon the type of metal being joined, the position of the weld being made, and the quality of the weld metal necessary for the application.

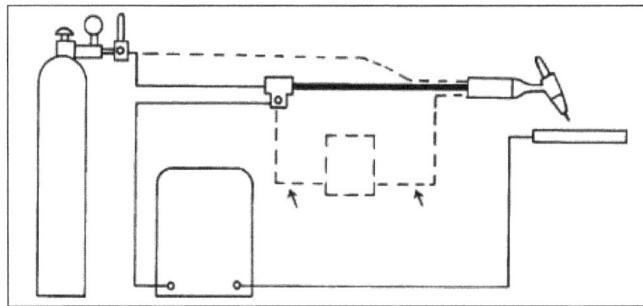

The basic equipment consists of the following:

- The power source
- Electrode holder (torch)
- Shielding gas
- Tungsten electrode
- Water supply when necessary
- Ground cable
- Protective equipment

Power Sources: Both AC and DC power sources are used in gas tungsten arc welding. They are the constant current type with a drooping volt-ampere curve. This type of power source produces very slight changes in the arc current when the arc length (voltage) is varied.

The choice between an AC or DC welder depends on the type and thickness of the metal to be welded. Distinct differences exist between AC and DC arc characteristics, and if DC is chosen, the polarity also becomes an important factor. The effects of polarity in GTAW are directly opposite the effects of polarity in SMAW. In SMAW, the distribution of heat between the electrode and work, which determines the penetration and weld bead width, is controlled mainly by the ingredients in the flux coating on the electrode. In GTAW where no flux coating exists, heat distribution between the electrode and the work is controlled solely by the polarity. The choice of the proper welding current will be better understood by analyzing each type separately.

Direct current electrode negative (DCEN) is produced when the electrode is connected to the negative terminal of the power source. Since the electrons flow from the electrode to the plate, approximately 70% of the heat of the arc is concentrated at the work, and approximately 30% at the electrode end. This allows the use of smaller tungsten electrodes that produce a relatively narrow concentrated arc. The weld shape has deep penetration and is quite narrow. Direct current electrode negative is suitable for welding most metals. Magnesium and aluminium have a refractory oxide coating on the surface that must be physically removed immediately prior to welding if DCSP is to be used.

Direct current electrode positive (DCEP) is produced when the electrode is connected to the positive terminal of the welding power source. In this condition, the electrons flow from the work to the electrode tip, concentrating approximately 70% of the heat of the arc at the electrode and 30% at the work. This higher heat at the electrode necessitates using larger diameter tungsten to prevent it from melting and contaminating the weld metal. Since the electrode diameter is larger and the heat is less concentrated at the work, the resultant weld bead is relatively wide and shallow.

Direct current electrode positive is rarely used in gas-tungsten arc welding. Despite the excellent oxide cleaning action, the lower heat input in the weld area makes it a slow process, and in metals having higher thermal conductivity, the heat is rapidly conducted away from the weld zone. When used, DCEP is restricted to welding thin sections (under 1/8") of magnesium and aluminium.

Alternating current is actually a combination of DCEN and DCEP and is widely used for welding aluminium. In a sense, the advantages of both DC processes are combined, and the weld bead produced is a compromise of the two. Remember that when welding with 60 Hz current, the electron flow from the electrode tip to the work reverses direction 120 times every second. Thereby, the intense heat alternates from electrode to work piece, allowing the use of an intermediate size electrode. The weld bead is a compromise having medium penetration and bead width. The gas ions blast the oxides from the surface of aluminium and magnesium during the positive half cycle.

DC constant current power sources - Constant current power sources, used for shielded metal arc welding, may also be used for gas-tungsten arc welding. In applications where weld integrity is not of utmost importance, these power sources will suffice. With machines of this type, the arc must be initiated by touching the tungsten electrode to the work and quickly withdrawing it to maintain the proper arc length. This starting method contaminates the electrode and blunts the point which has been grounded on the electrode end. These conditions can cause weld metal inclusions and poor arc direction. Using a power source designed for gas tungsten arc welding with a high frequency stabilizer will eliminate this problem. The electrode need not be touched to the work for arc initiation. Instead, the high frequency voltage, at very low current, is superimposed onto the welding current. When the electrode is brought to within approximately 1/8 inch of the base metal, the high frequency ionizes the gas path, making it conductive and a welding arc is established.

The high frequency is automatically turned off immediately after arc initiation when using direct current.

AC Constant Current Power Source - Designed for gas tungsten arc welding, always incorporates high frequency, and it is turned on throughout the weld cycle to maintain a stable arc. When welding with AC, the current passes through 0 twice in every cycle and the must be reestablished each time it does so. The oxide coating on metals, such as aluminium and magnesium, can act much like a rectifier. The positive half-cycle will be eliminated if the arc does not reignite, causing an unstable condition. Continuous high frequency maintains an ionized path for the welding arc, and assures arc reignition each time the current changes direction. AC is extensively used for welding aluminium and magnesium.

AC/DC Constant Current Power Sources - Designed for gas tungsten arc welding, are available, and can be used for welding practically all metals. The gas tungsten arc welding process is usually chosen because of the high quality welds it can produce. The metals that are commonly welded with this process, such as stainless steel, aluminium and some of the more exotic metals, cost many times the price of mild steel; and therefore, the power sources designed for this process have many desirable features to insure high quality welds. Among these are:

- Remote current control, which allows the operator to control welding amperage with a hand control on the torch, or a foot control at the welding station.

- Automatic soft-start, which prevents a high current surge when the arc is initiated.

- Shielding gas and cooling water solenoid valves, which automatically control flow before, during and for an adjustable length of time after the weld is completed.

- Spot-weld timers, which automatically control all elements during each spot-weld cycle. Other options and accessories are also available.

Power sources for automatic welding with complete programmable output are also available. Such units are used extensively for the automatic welding of pipe in position. The welding current is automatically varied as the torch travels around the pipe. Some units provide a pulsed welding current where the amperage is automatically varied between a low and high several times per second. This produces welds with good penetration and improved weld bead shape.

Torches: The torch is actually an electrode holder that supplies welding current to the tungsten electrode, and an inert gas shield to the arc zone. The electrode is held in a collet-like clamping device that allows adjustment so that the proper length of electrode pro- trudes beyond the shielding gas cup. Manual torches are designed to accept electrodes of 3 inch or 7 inch lengths. Torches may be either air or water-cooled. The air-cooled types actually are cooled to a degree by the shielding gas that is fed to the torch head through a composite cable. The gas actually surrounds the copper welding cable, affording some degree of cooling. Water-cooled torches are usually used for applications where the welding current exceeds 200 amperes. The water inlet hose is connected to the torch head. Circulating around the torch head, the water leaves the torch via the current-in hose and cable assembly. Cooling the welding cable in this manner allows the use of a smaller diameter cable that is more flexible and lighter in weight.

The gas nozzles are made of ceramic materials and are available in various sizes and shapes. In some heavy duty, high current applications, metal water-cooled nozzles are used.

A switch on the torch is used to energize the electrode with welding current and start the shielding gas flow. High frequency current and water flow are also initiated by this switch if the power source is so equipped. In many installations, these functions are initiated by a foot control that also is capable of controlling the welding current. This method gives the operator full control of the arc. The usual welding method is to start the arc at a low current, gradually increase the current until a molten pool is achieved, and welding begins. At the end of the weld, current is slowly decreases and the arc extinguished, preventing the crater that forms at the end of the weld when the arc is broken abruptly.

Shielding Gases - Argon and helium are the major shielding gases used in gas tungsten arc welding. In some applications, mixtures of the two gases prove advantageous. To a lesser extent, hydrogen is mixed with argon or helium for special applications.

Argon and helium are colorless, odorless, tasteless and nontoxic gases. Both are inert gases, which means that they do not readily combine with other elements. They will not burn nor support combustion. Commercial grades used for welding are 99.99% pure. Argon is .38% heavier than air and about 10 times heavier than helium. Both gases ionize when present in an electric arc. This means that the gas atoms lose some of their electrons that have a negative charge. These unbalanced gas atoms, properly called positive ions, now have a positive charge and are attracted to the negative pole in the arc. When the arc is positive and the work is negative, these positive ions impinge upon the work and remove surface oxides or scale in the weld area.

Argon is most commonly used of the shielding gases. Excellent arc starting and ease of use make it most desirable for manual welding. Argon produces a better cleaning action when welding aluminium and magnesium with alternating current. The arc produced is relatively narrow. Argon is more suitable for welding thinner material. At equal amperage, helium produces a higher arc voltage than argon. Since welding heat is the product of volts times amperes, helium produces more available heat at the arc. This makes it more suitable for welding heavy sections of metal that have high heat conductivity, or for automatic welding operations where higher welding speeds are required.

Argon-helium gas mixtures are used in applications where higher heat input and the desirable characteristics of argon are required. Argon, being a relatively heavy gas, blankets the weld area at lower flow rates. Argon is preferred for many applications because it costs less than helium. Helium, being approximately 10 times lighter than argon, requires flow rates of 2 to 3 times that of argon to satisfactorily shield the arc.

Electrodes - Electrodes for gas tungsten arc welding are available in diameters from .010" to 1/4" in diameter and standard lengths range from 3" to 24". The most commonly used sizes, however, are the .040", 1/16", 3/32", and 1/8" diameters.

The shape of the tip of the electrode is an important factor in gas tungsten arc welding. When welding with DCEN, the tip must be ground to a point. The included angle at which the tip is ground varies with the application, the electrode diameter, and the welding current. Narrow joints require a relatively small included angle. When welding very thin material at low currents, a needlelike point ground onto the smallest available electrode may be necessary to stabilize the arc. Properly ground electrodes will assure easy arc starting, good arc stability, and proper bead width.

When welding with AC, grinding the electrode tip is not necessary. When proper welding current is used, the electrode will form a hemispherical end. If the proper welding current is exceeded, the end will become bulbous in shape and possibly melt off to contaminate the weld metal.

The American Welding Society has published Specification AWS A5.12-80 for tungsten arc welding electrodes that classifies the electrodes on the basis of their chemical composition, size and finish. Briefly, the types specified are listed below:

- Pure Tungsten (AWS EWP) Color Code: Green Used for less critical applications. The cost is low and they give good results at relatively low currents on a variety of metals. Most stable arc when used on AC, either balanced wave or continuous high frequency.

- 1% Thoriated Tungsten (AWS EWTh-1) Color Code: Yellow Good current carrying capacity, easy arc starting and provide a stable arc. Less susceptible to contamination. Designed for DC applications of nonferrous materials.

- 2% Thoriated Tungsten (AWS EWTh-2) Color Code: Red Longer life than 1% Thoriated electrodes. Maintain the pointed end longer, used for light gauge critical welds in aircraft work. Like 1%, designed for DC applications for nonferrous materials.

- 5% Thoriated Tungsten (AWS EWTh-3) Color Code: Blue Sometimes called "striped" electrode because it has 1.0-2.0% Thoria inserted in a wedge-shaped groove throughout its length. Combines the good properties of pure and thoriated electrodes. Can be used on either AC or DC applications.

- Zirconia Tungsten (AWS EWZr) Color Code: Brown Longer life than pure tungsten. Better performance when welding with AC. Melts more easily than thoriam-tungsten when forming rounded or tapered tungsten end. Ideal for applications where tungsten contamination must be minimized.

Shielded Metal Arc Welding

Shielded Metal Arc Welding (SMAW) is an electric arc welding process in which an electric arc between a covered metal electrode and the work generates the heat for welding. The filler metal is deposited from the electrode, and the electrode covering provides the shielding. Some slang names for this process are "stick welding" or "stick electrode welding".

The shielded metal arc welding process is one of the simplest and most versatile arc welding processes. It can be used to weld both ferrous and non-ferrous metals, and it can weld thicknesses above approximately 18 gauge in all positions. The arc is under the control of the welder and is visible. The welding process leaves slag on the surface of the weld bead which must be removed. The most common use for this process is welding mild and low alloy steels. The equipment is extremely rugged and simple, and the process is flexible in that the welder needs to take only the electrode holder and work lead to the point of welding.

Most sources give credit for the invention of the electric arc to Sir Humphrey Davy of England in 180l. For the most part, the electric arc remained a scientific novelty until 1881, when the carbon arc street lamp was invented and the first attempts to weld using the carbon arc process were made. The metal arc welding process came into being when metal rods replaced the carbon electrodes in 1889. Coverings for the bare wire electrodes were first developed in the early 1900's. The first major use occurred during World War I, especially in the shipbuilding industry. After the war, there was a period of slow growth until the early 1930's when shielded metal arc welding became a major manufacturing method and a dominant welding process. Today, the shielded metal arc welding process is widely used, even though its relative importance has been declining slowly in recent years.

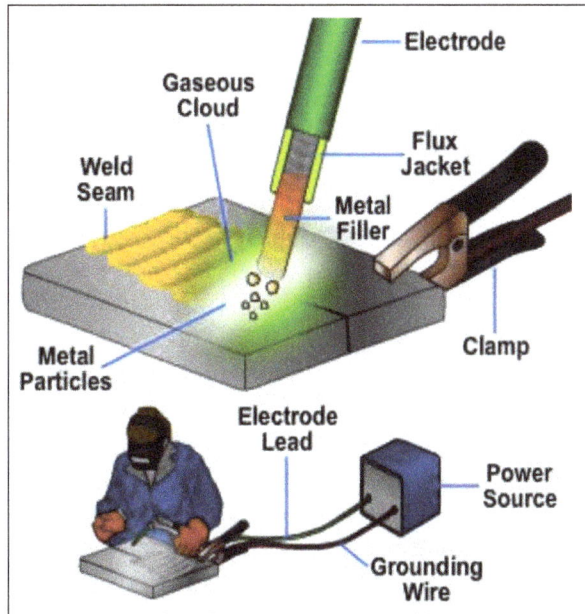

Shielded metal arc welding.

Methods of Application

The shielded metal arc welding process is basically a manually operated process. The electrode is clamped in an electrode holder and the welder manipulates the tip of the electrode in relation to the metal being welded. The welder strikes, maintains, and stops the arc manually. Several variations of this process are done automatically rather than manually. These are: gravity welding, firecracker welding, and massive electrode welding. These methods comprise only a very small percentage of welding done by the shielded metal arc welding process.

Advantages and Limitations

Shielded metal arc welding is widely used because of its versatility, portability, and comparatively simple and inexpensive equipment. In addition, it does not require auxiliary gas shielding or granular flux. Welders can use the shielded metal arc welding process for making welds in any position they can reach with an electrode. Electrodes can be bent so they can be used to weld blind areas. Long leads can be used to weld in many locations at great distances from the power source. Shielded metal arc welding can be used in the field because the equipment is relatively light and portable. This process is also less sensitive to wind and draft than gas shielded arc welding processes. Shielded metal arc welding can be used to weld a wide variety of metal thicknesses. This process is more useful than other welding processes for welding complex structural assemblies because it is easier to use in difficult locations and for multi-position welding. Shielded metal arc welding is also a popular process for pipe welding because it can create weld joints with high quality and strength. However, the shielded metal arc welding process has several limitations. Operator duty cycle and overall deposition rates for covered electrodes are usually less than those of a continuous electrode process. This is because electrodes have a fixed length and welding must stop after each electrode has been consumed to discard the remaining portion of the used electrode clamped into the holder and reapply another. Another limitation is that the slag must be removed from the weld after every pass. Finally, the shielded metal arc welding process cannot be used to weld some of the non-ferrous metals.

Principles of Operation

The shielded metal arc welding process uses the heat of the electric arc to melt the consumable electrode and the work being welded. The welding circuit includes a power source, welding cables, an electrode holder, a work clamp and a welding electrode. One of the welding cables connects the power source to the electrode holder and the other cable connects to the workpiece.

The welding begins when the welder initiates the arc by momentarily touching the electrode to the base metal, which completes the electrical circuit. The welder guides the electrode manually, controlling both the travel speed and the direction of travel. The welder maintains the arc by controlling the distance between the work material and the tip of the electrode (length of the arc). Some types of electrodes can be dragged along the surface of the work so that the coating thickness controls the arc length, which controls the voltage.

The heat of the arc melts the surface of the base metal and forms a molten weld puddle. The melted electrode metal is transferred across the arc and becomes the deposited weld metal. The deposit is covered by a slag produced by components in the electrode coating. The arc is enveloped in a gas shield provided by the disintegration of some of the ingredients of the electrode coating. Most of the electrode core wire is transferred across the arc, but small particles escape from the weld area as spatter, and a very small portion leaves the welding area as smoke.

Arc Systems

The constant current type of power source is best for shielded metal arc welding. The constant current welding machines provide a nearly constant welding current for the arc.

The constant current output is obtained with a drooping volt ampere characteristic, which means that the voltage reduces as the current increases. The changing arc length causes the arc voltage to increase or decrease slightly, which in turn changes the welding current. Within the welding range, the steeper the slope of the volt-ampere curve, the smaller the current change for a given change in the arc voltage.

Under certain conditions, there is a need for variations in the volt-ampere slope. A steep volt-ampere characteristic is desirable when the welder wants to achieve maximum welding speed on some welding jobs. The steeper slope gives less current variation with changing arc length, and it gives a softer arc. The types of machines that have this kind of curve are especially useful on sheet metals. Machines with this characteristic are typically used with large diameter electrodes and high amperages. On some applications, such as welding over rust, or a position pipe welding where better arc control with high penetration capability is desired, a less steep volt-ampere characteristic is more desirable. Machines with the less steep volt-ampere curve are also easier to use for depositing the root passes on joints with varying fit-up. This type of power source characteristic allows the welder to control the welding current in a specific range by changing the arc length and producing a more driving arc. Differences in the basic power source designs cause these variations in the power sources. Figure shows volt-ampere curves for different performance characteristics. This shows several slopes, all of which can provide the same normal voltage and current. The flatter slopes give a greater current variation for a given voltage change or arc length change. Machines that have a higher short circuit current give more positive starting.

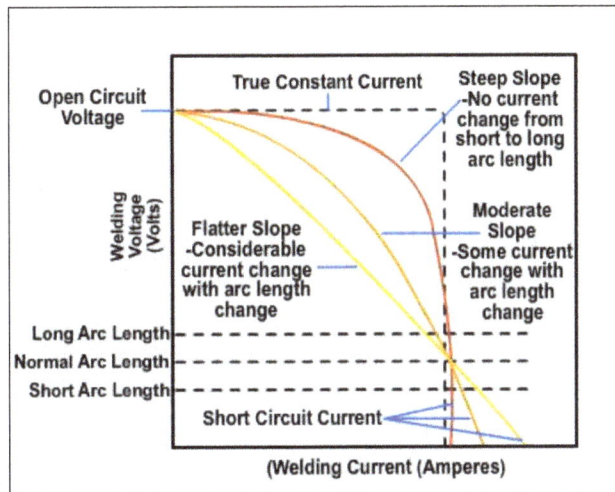

Typical volt-ampere curves for constant current types of power sources.

Electrical Terms

Many terms are associated with arc welding. The following basic terms are especially important.

- Alternating current: Alternating current is an electrical current that has alternating negative and positive values. In the first half-cycle, the current flows in one direction and then reverses itself for the next half-cycle. In one complete cycle, the current spends 50 percent of the time flowing one way and the other 50 percent flowing the other way. The rate of change in direction is called frequency, and it is indicated by cycles per second. In the United States, the alternating current is set at 60 cycles per second.

- Ampere: Amperes, sometimes called "amps," refers to the amount of current that flows through a circuit. It is measured by an "amp" meter.

- Conductor: Conductor means any material that allows the passage of an electrical current. Current: Current is the movement or flow of an electrical charge through a conductor.

- Direct current: Direct current is an electrical current that flows in one direction only.

- Electrical circuit: Electrical circuit is the path taken by an electrical current flowing through a conductor from one terminal of the source to the load and returning to the other terminal of the source.

- Polarity: Polarity is the direction of the flow of current in a circuit. Since current flows in one direction only in a dc welder, the polarity becomes an important factor in welding operations.

- Resistance: Resistance is the opposition of the conductor to the flow of current. Resistance causes electrical energy to be changed into heat.

- Volt: A volt is the force required to make the current flow in an electrical circuit. It can be compared to pressure in a hydraulic system. Volts are measured with a volt meter.

Metal Transfer

The intense heat of the welding arc melts the tip of the electrode and melts the surface of base metal. The temperature of the arc is about 9000 °F (5000 °C) which causes almost instantaneous melting of the surface of the work. Globules form on the tip of the electrode and transfer through the arc to the molten weld puddle on the surface of the work. When the detaching globules are small during the transfer, this is known as spray type metal transfer. When the globules are relatively large during transfer, it is known as globular type metal transfer. Surface tension sometimes causes a globule of metal to connect the tip of the electrode to the weld puddle. This causes an electrical short and makes the arc go out. Usually this is a momentary occurrence, but occasionally the electrode will stick to the weld puddle. When the short circuit occurs, the current builds up to a short circuit value and the increased current usually melts the connecting metal and reestablishes the arc. A welding machine with a flatter volt-ampere curve will give a higher short circuit current than one with a steeper volt-ampere curve. The electrode sticking problem will be slightly less with a machine that has a flatter volt-ampere curve. A softer arc, produced by a steeper slope, will decrease the amount of weld spatter. A more driving arc, produced by a flatter slope, causes a more violent transfer of metal into the weld puddle, which will cause a greater splashing effect. This greater splashing effect will generate more spattering from the weld puddle. When the welds are made in the flat or horizontal positions, the forces of gravity, magnetism, and surface tension induce the transfer of the metal. When the welds are made in the vertical or overhead positions, the forces of magnetism and surface tension induce the metal transfer, while the force of gravity opposes metal transfer. Lower currents are used for vertical and overhead welding to allow shorter arc lengths and promote a smaller metal droplet size less affected by gravity.

Equipment for Welding

The equipment for the shielded metal arc welding process consists of a power source, welding cable, electrode holder, and work clamp or attachment. It shows a figure of the equipment.

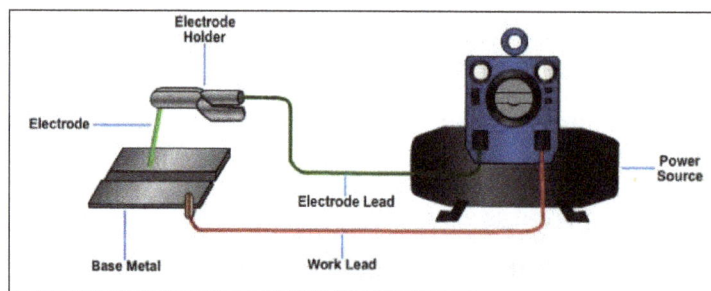

Equipment for shielded metal arc welding.

Power Sources

The purpose of the power source or welding machine is to provide the electric power of the proper current and voltage to maintain a welding arc. Many different sizes and types of power sources are designed for shielded metal arc welding. Most power sources operate on 230 or 460 volt input electric power, but power sources that operate on 200 or 575 volt input power are also available.

Types of Current

Shielded metal arc welding can use either direct current (DC) or alternating current (AC). Electrode negative (straight polarity) or electrode positive (reverse polarity) can be used with direct current. Each type of current has distinct advantages, but selection of the type of welding current used, usually depends on the availability of equipment and the type of electrode selected. Direct current flows in one direction continuously through the welding circuit.

The advantages it has over alternating current are: Direct current is better at low currents and with small diameter electrodes.

All classes of covered electrodes can produce satisfactory results. Arc starting is generally easier with direct current. Maintaining a short arc is easier.

Direct current is easier to use for out-of position welding because lower currents can be used. Direct current is easier to use for welding sheet metal. It generally produces less weld spatter than alternating current.

Polarity or direction of current flow is important in the use of direct current. Electrode negative (straight polarity) is often used when shallower penetration is required. Electrode positive (reverse polarity) is generally used where deep penetration is needed. Normally, electrode negative provides higher deposition rates than electrode positive. The type of electrode often governs the polarity to be used.

Alternating current is a combination of both polarities that alternates in regular cycles. In each cycle the current starts at zero, builds up to a maximum value in one direction, decays back to zero, builds up to a maximum value in the other direction, and again decays to zero. The polarity of the alternating current changes 120 times during the 60 Hertz cycle used in the United States. Depths of penetration and deposition rates for alternating current are generally intermediate between those for DC electrode positive and DC electrode negative. Some advantages of alternating current are: Arc blow is rarely a problem with alternating current.

Alternating current is well suited for welding thick sections using large diameter electrodes.

Power Source Duty Cycle

Duty cycle is the ratio of arc time to total time. For a welding machine, a 10 minute time period is used. Thus, for a 60% duty cycle machine, the welding load would be applied continuously for 6 minutes and would be off for 4 minutes. Most industrial type constant current machines are rated at 60% duty cycle. The formula for determining the duty cycle of a welding machine for a given load current is:

$$\% \text{Duty Cycle} = \frac{(\text{Rated Current})^2}{(\text{Load Current})^2} \times \text{Rated Duty Cycle}$$

For example, if a welding machine is rated at a 60% duty cycle at 300 amperes, the duty cycle of the machine when operated at 350 amperes would represent the ratio of the square of the rated current to the square of the load current multiplied by the rated duty cycle. A line is drawn parallel to the sloping lines through the intersection of the subject machines rated current output and rated duty cycle. For example, a question might arise whether a 400 amp 60% duty cycle machine could be used for a fully automatic requirement of 300 amps for a 10-minute welding job. It shows that the machine can be used at slightly over 300 amperes at a 100% duty cycle. Conversely, there may be a need to draw more than the rated current from a welding machine, but for a short period. This illustration can be used to compare various machines. Relate all machines to the same duty cycle for a time comparison:

$$\%\,Duty\,Cycle = \frac{(300)^2}{(350)^2} \times 60 = 44\%$$

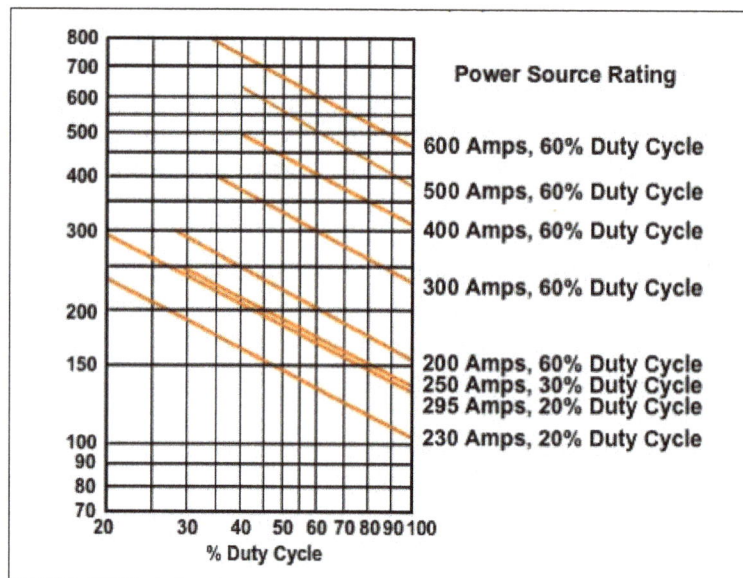

Duty cycle vs. current load.

Types of Power Sources

The output characteristics of the power source must be of the constant-current (CC) type. The normal current range is 25 to 500 amps using conventional size electrodes. The arc voltage is 15 to 35 volts.

Generator and Alternator Welding Machines

The generator can be powered by an electric motor for shop use or by an internal combustion engine (gasoline, gas, or diesel) for field use. Engine driven welders can have either water or air cooled engines, and many of them provide auxiliary power for emergency lighting, power tools, etc. Generator welding machines can provide both AC and DC power.

An alternator welding machine is an electric generator that produces AC power. This power source has a rotating assembly.

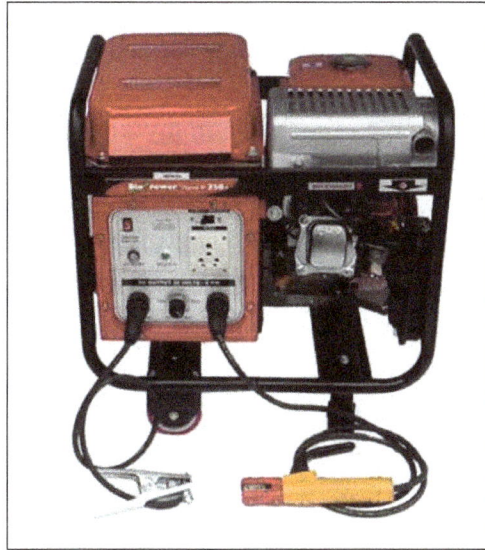
Portable welder/generator.

These machines are also called rotating or revolving field machines.

On dual control machines, normally a generator, the slope of the output curve can vary. The fine adjustment control knob controls open circuit, or "no load", voltage. This control is also the fine welding current adjustment during welding. The range switch provides coarse adjustment of the welding current. In this way, a soft or harsh arc can be obtained. With the flatter curve and its low open circuit voltage, a change in arc voltage will produce a greater change in output current. This produces the digging arc preferred for pipe welding. With a steeper curve and its high open circuit voltage, the same change in arc voltage will produce less of a change in output current. This is a soft or quiet arc, useful for sheet metal welding. This type of welding machine gives the smoothest operating arc because it produces less voltage ripple.

Diesel engine driven power source.

Transformer Welding Machines

The transformer type welding machine is the least expensive, lightest, and smallest type of welder.

It produces alternating current for welding. The transformer welder takes power directly from the line, transforms it to the power required for welding, and by means of various magnetic circuits, inductors, etc., provides the volt-ampere characteristics proper for welding. The welding current output of a transformer welder may be adjusted in many different ways. The simplest method of adjusting output current is to use a tapped secondary coil on the transformer. This is a popular method many of the limited input, small welding transformers employ. The leads to the electrode holder and the work are connected to plugs, which the welder may be insert in sockets on the front of the machine in various locations to provide the required welding current. Some machines employ a tap switch instead of the plug-in arrangement. In any case, exact current adjustment is not entirely possible.

Industrial types of transformer welders usually employ a continuous output current control. This can be obtained by mechanical means, or electrical means. The mechanical method usually involves moving the core of the transformer. Any method that involves mechanical movement of the transformer parts requires considerable movement for full range adjustment. The more advanced method of adjusting current output is by means of electrical circuits. In this method the core of the transformer or reactor is saturated by an auxiliary electric circuit which controls the amount of current delivered to the output terminals. By merely adjusting a small knob, to the welder can provide continuous current adjustment from the minimum to maximum of the output.

Although the transformer type of welder has many desirable characteristics, it also has some limitations. The power required for a transformer welder must be supplied by a single phase system, and this may create an unbalance of the power supply lines, which is objectionable to most power companies. In addition, transformer welders have a rather low power factor unless they are equipped with power factor correcting capacitors. The addition of capacitors corrects the power factor under load and Diesel engine driven power source. produces a reasonable power factor that is not objectionable to electric power companies.

Transformer welders have the lowest initial cost. They require less space and are normally quiet in operation. In addition, alternating current welding power supplied by transformers reduces arc blow, which can be troublesome on many welding applications. They do not, however, have as much flexibility for the operator as the dual controlled generator.

Transformer-Rectifier Welding Machines

The previously described transformer welders provide alternating current to the arc. Some types of electrodes operate successfully only with direct current power. A method of supplying direct current power to the arc without using a rotating generator is adding a rectifier, an electrical device which changes alternating current into direct current. Transformer-rectifier welding machines operate on single phase input power. These machines are used when both AC and DC current are needed. A single phase type of AC welder is connected to the rectifier which then produces DC current for the arc. By means of a switch that can change the output terminals to the transformer or the rectifier, the operator can select either AC or DC current for the welding requirement. Transformer-rectifier welding machines are available in different sizes. These machines are more efficient electrically than the generator welding machines, and they provide quieter operation. Figure shows an AC/DC single phase power source.

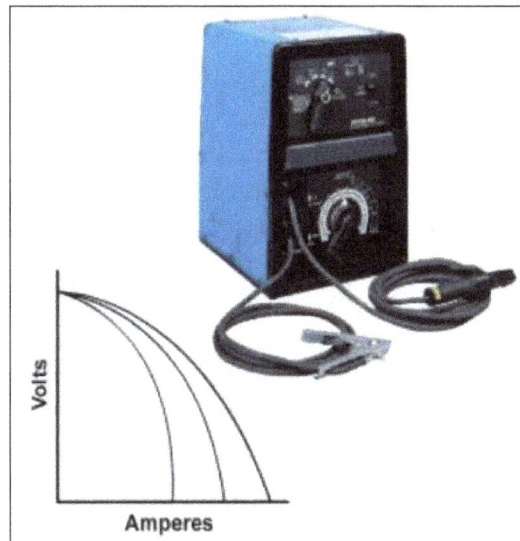

AC/DC single phase power source.

Three Phase Rectifier Welding Machines

Three phase rectifier welding machines provide DC welding current to the arc. These machines operate on three phase input power. The three phase input helps overcome the line unbalance that occurs with single phase transformer-rectifier welding machines. In this type of machine, the transformers feed into a rectifier bridge, which then produces direct current for the arc. The three-phase rectifier unit is more efficient electrically than a generator and provides quiet operation. This type of machine also gives the least voltage ripple and produces the smoothest arc of the static type welding machines. Figure shows a three phase solid state constant voltage power source. It automatically monitors output voltage and makes required changes to compensate for line voltage fluctuation.

Multiple Operator System

A multiple operator welding system uses a heavy duty, high current, and relatively high voltage power source which feeds a number of individual operator welding stations. At each welding station, a variable resistance is adjusted to drop the current to the proper welding range. Based on the duty cycle of the welding equipment, one welding machine can supply welding power simultaneously to a number of welding operators. The current supplied at the individual station has a drooping characteristic similar to the single operator welding machines described above. The power source, however, has a constant voltage Output. Constant voltage power sources are those that maintain a constant voltage for a given current setting. The volt-ampere curve for this type of power source is nearly flat. The welding machine size and the number and size of the individual welding current control stations must be carefully matched for an efficient multiple operator system. The formula for determining the number of arcs that can be operated off of one power source is:

$$\frac{\text{Available Power}}{\text{Average Arc Amperes} \times \text{Duty Cycle}} = \text{Number of Arcs}$$

Inverter Power Sources

In this type of power source, which utilizes the inverter, the power from the line is first rectified to pulsing direct current. This current then goes to a high frequency oscillator or chopper, which changes the DC into high-voltage, highfrequency AC in the range 5 to 30 kHz. The output of the chopper circuit is controlled in accordance with welding procedure requirements. The high frequency AC is then transformed down to the operating welding voltage. The advantage of the inverter is the use of a small lightweight transformer, since transformers become smaller as frequency increases. The high frequency AC current is then rectified with silicon diodes to provide direct current output at normal welding current and voltage. The inverter power source has become economically feasible due to the availability of high current, high speed solid state electronic components at a reasonable cost. Inverter power sources are about 25% the weight of a conventional rectifier of the same power capacity and about 33% of the size. They provide higher electrical efficiency, a higher power factor, and a faster response time. Several variations of the inverter power source are available.

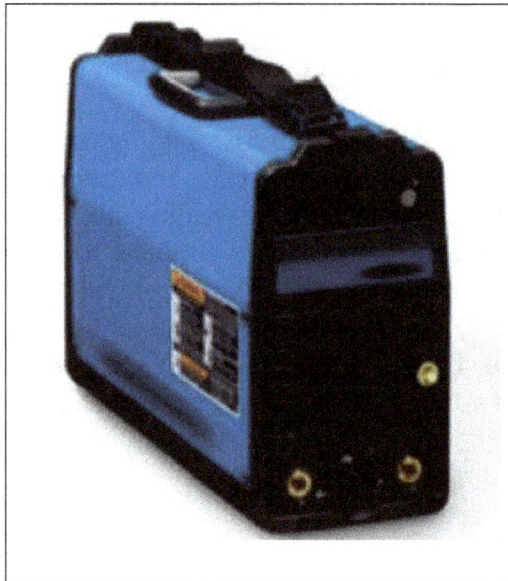

Inverter power source.

Selecting a Power Source

- Selecting a welding machine is based on the amount of current required for the work.

- The power available to the job site. Convenience and economic factors.

- The size of the machine is based on the welding current and duty cycle required. Welding current, duty cycle, and voltage are determined by considering weld joints, weld sizes, and welding procedures. The incoming power available dictates this fact. Finally, the job situation, personal preference, and economic considerations narrow the field to the final selection. Consult the local welding equipment supplier to help make your selection. Know the following data when selecting a welding power source: Rated load amperes (current) Duty cycle Voltage of power supply (incoming) Frequency of power supply (incoming) Number of phases of power supply (incoming).

Controls

The controls are usually located on the front panel of the welding machine. These usually consist of a knob or tap switch to set the rough current range and a knob to adjust the current within the set range. On DC welding machines there is usually a switch to change polarity, and on combination AC-DC machines, there is usually a switch to select the polarity or AC current. An On-Off switch is also located on the front of the machine.

Arc Force Control is a function of amperage triggered by a preset (internal module) voltage. The preset trigger voltage is 18 volts. What this means is that anytime the arc voltage drops from normal welding voltage to 18 volts or less, the drop triggers the arc force current, which gives the arc a surge of current to keep the arc from going out.

When an arc is struck, the electrode is scratched against the work. At that point, the voltage goes to -0- which triggers the arc force current and the arc is initiated quickly. On a standard machine without arc force control, arc striking is difficult and electrode sticking may occur.

After the arc is established, a steady burn-off is desired. As the electrode burns and droplets of metal are transferred from the end of the electrode to the work piece, there is a time period when the droplet is still connected to the end of the electrode but is also touching the work piece. When this occurs, the machine is, in effect, in a "dead-short" - the voltage drops, the arc force is triggered and the droplet is transferred. On machines without arc force, the burn-off is the same; however, without the arc force to help, an arc outage may occur, and the electrode will stick in the puddle.

In tight joints, such as pipe welding, the arc length is very short and with standard machines, it is difficult to maintain the arc since it wants to "short-out" against the sidewalls or bottom of the joint. The arc force control can be adjusted on this type application to prevent electrode sticking; whenever the voltage drops, the drop triggers the arc force current and the sticking doesn't happen because the current surge occurs.

In many applications, there is a need for a very forceful arc to obtain deeper penetration, or in the case of arc gouging, the forceful arc is essential in helping to force the metal out of the groove being gouged. With arc force control, this type application is made much easier than with conventional machines, with which arc length becomes critical and arc outages can occur.

When welding with a given size electrode, there is always an optimum amperage setting. When using arc force control, the optimum amperage setting is continually working to maintain the arc, which means that although we can't see it on the meters, there is usually some added amperage to assist in rod burn-off. This in turn means we really get a slightly faster burn-off than with a conventional rectifier.

When working out-of-position, a forceful arc is needed to help put metal in place. Each individual operator can adjust the arc force control to provide just the amount needed. Arc force can also be of assistance when welding rusty or scaly material, since the more forceful arc will help to break up these deposits.

Electrode Holder

An electrode holder, commonly called a stinger, is a clamping device for holding the electrode securely in any position. The welding cable attaches to the holder through the hollow insulated handle. The design of the electrode holder permits quick and easy electrode exchange. Two general types of electrode holders are in use: insulated and noninsulated. The noninsulated holders are not recommended because they are subject to accidental short circuiting if bumped against the workpiece during welding. For safety reasons, try to ensure the use of only insulated stingers on the jobsite.

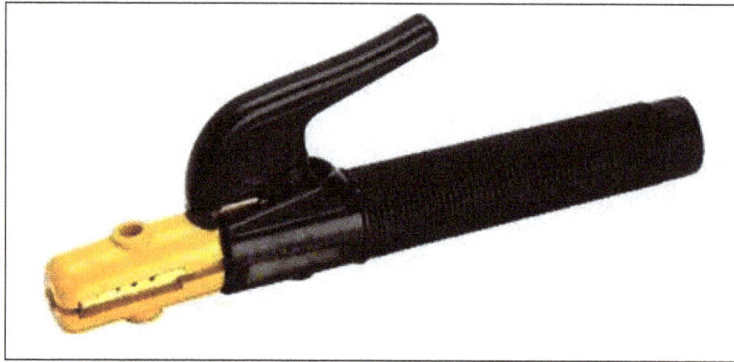

Insulated pincher and collet types of electrode holders.

Electrode holders are made in different sizes, and each manufacturer has its own system of designation. Each holder is designed for use within a specified range of electrode diameters and welding current. Welding with a machine having a 300-ampere rating requires a larger holder than welding with a 100-ampere machine. If the holder is too small, it will overheat.

Welding Cables

The welding cables and connectors connect the power source to the electrode holder and to the work. These cables are normally made of copper or aluminium. The cable that connects the work to the power source is called the work lead. The work leads are usually connected to the work by pincher clamps or a bolt. The cable that connects the electrode holder to the power source is called the electrode lead.

The welding cables must be flexible, durable, well insulated, and large enough to carry the required current. Use only cable specifically designed for welding. Always use a highly flexible cable for the electrode holder connection. This is necessary so the operator can easily maneuver the electrode holder during the welding process. The work lead cable need not be so flexible because once it is connected, it does not move.

Two factors determine the size of welding cable to use: the amperage rating of the machine and the distance between the work and the machine. If either amperage or distance increases, the cable size must also increase. Cable sizes range from the smallest at AWG No.8 to AWG No. 4/0 with amperage ratings of 75 amperes and upward. recommended cable sizes for use with different welding currents and cable lengths. The best size cable is one that meets the amperage demand but is small enough to manipulate easily.

As a rule, the cable between the machine and the work should be as short as possible. Use one continuous length of cable if the distance is less than 35 feet. If you must use more than one length of cable, join the sections with insulated lock-type cable connectors. Joints in the cable should be at least 10 feet away from the operator.

Table: Suggested copper welding cable sizes for SMAW.

Weld Type	Welding Current	Length of Cable Curcuit in Feet – Cable Size A.W.G.					
		60'	100'	150'	200'	300'	400'
Manual (Low Duty Cycle)	100	4	4	4	2	1	1/0
	150	2	2	2	1	2/0	2/0
	200	2	2	1	1/0	3/0	3/0
	250	2	2	1/0	2/0		
	300	1	1	2/0	3/0		
	350	1/0	1/0	3/0	4/0		
	400	1/0	1/0	3/0			
	450	2/0	2/0	4/0			
	500	2/0	2/0	4/0			

Ground Clamps

A good ground clamp is essential to produce quality welds. Without proper grounding, the circuit voltage fails to produce enough heat for proper welding, and there is the possibility of damage to the welding machine and cables. Three basic methods are used to ground a welding machine. You can fasten the ground cable to the workbench with a C-clamp, attach a spring-loaded clamp directly onto the workpiece, or bolt or tack-weld the end of the ground cable to the welding bench. The third way creates a permanent common ground.

Accessories

Accessory equipment used for shielded metal arc welding consists of items used for removing slag and cleaning the weld bead. Chipping hammers are often used to remove the slag. Wire brushes or grinders are the most common methods for cleaning the weld.

Manufacturers offer various options and accessories also, depending on the type of power source and the procedure recommendations.

Equipment Operation and Maintenance

Learning to arc weld requires many skills. Among these are the abilities to set up, operate, and maintain your welding equipment.

In most factory environments, the work is brought to the welder. In the Seabees, the majority of the time the opposite is true. You will be called to the field for welding on buildings, earthmoving equipment, well drilling pipe, ship to shore fuel lines, pontoon causeways, and the list goes on. To accomplish these tasks, you have to become familiar with your equipment and be able to maintain it in the field. It would be impossible to give detailed maintenance information here because of the many different types of equipment found in the field; therefore, we will only cover the highlights.

Become familiar with the welding machine you will be using. Study the manufacturer's literature and check with your senior petty officer or chief on items you do not understand. Machine setup involves selecting current type, polarity, and current settings. The current selection depends on the size and type of electrode used, position of the weld, and the properties of the base metal.

Cable size and connections are determined by the distance required to reach the work, the size of the machine, and the amperage needed for the weld.

Operator maintenance depends on the type of welding machine used. Transformers and rectifiers require little maintenance compared to enginedriven welding machines. Transformer welders require only to be kept dry and need a minimal amount of cleaning. Only electricians should perform internal maintenance due to the possibility of electrical shock. Enginedriven machines require daily maintenance of the motors. In most places you will be required to fill out and turn in a daily inspection form called a "hard card" before starting the engine. This form is a list of items, such as oil level, water level, visible leaks, and other things, that affect the operation of the machine.

After checking all of the above items, you are now ready to start welding. Listed below are some additional welding rules you must follow: Clear the welding area of all debris and clutter. Do not use gloves or clothing that contain oil or grease. Check that all wiring and cables are installed properly. Ensure that the machine is grounded and dry. Follow all the manufacturer's directions on operating the welding machine. Have on-hand a protective screen to protect others in the welding area from flash burns.

Always keep fire-fighting equipment on hand. Clean rust, scale, paint, or dirt from the joints to be welded.

Covered Electrodes

The covered electrode provides both the filler metal and the shielding for the shielded metal arc welding process. Covered electrodes have different compositions of core wire and a wide variety of types of flux coverings that perform one or all of the following functions, depending upon the type of electrode:

- Forming a slag blanket over the molten puddle and solidified weld.

- Providing shielding gas to prevent atmospheric contamination of both the arc stream and the weld metal.

- Providing ionizing elements for smoother arc operation.

- Provides deoxidizers and scavengers to refine the grain structure of the weld metal.

- Providing alloying elements such as nickel and chromium for stainless steel.

- Providing metal such as iron powder for higher deposition rates.

The first two functions listed prevent the pickup of nitrogen and oxygen into the weld puddle and the red hot solidified weld metal. The nitrogen and oxygen form nitrides and oxides which cause the weld metal to become brittle.

Welding Applications

Shielded metal arc welding is widely used because of its versatility. Welding can be performed at a distance from the power source which makes it popular for welding in the field. The equipment for this process is relatively simple to operate, portable, and inexpensive. Shielded metal arc welding is a major process used for maintenance and repair work. It is popular in small production shops where limited capital is available and where the amount of welding done is minor compared to other manufacturing operations. Shielded metal arc welding is often used for tacking parts together which are then welded by another process.

Industries

Shielded metal arc welding is the welding process of choice in a number of civilian industries because it is versatile and user friendly. It has been replaced in recent years by flux cored arc welding but remains competitive because of the low equipment costs and wide applicability.

Field Welded Storage Tanks

Field welded storage tanks differ from pressure vessels because they are used to store petroleum, water or other liquids at atmospheric pressure. Shielded metal arc welding is widely used in the fabrication and erection of field welded storage tanks. These tanks are generally constructed of low-carbon and structural steels. Nickel steels are employed when the tanks require higher toughness. This process is used to weld longitudinal and circumferential seams on the tanks as well as the structural support members. Figure shows field welding of a large circumference pipe. An engine driven generator power source is being used because there is no electricity available.

Pipe welding.

Pressure Vessels

Pressure vessels and boilers are also welded using this process. Shielded metal arc welding is primarily used for welding attachments to the vessel. This kind of welding commonly uses all sizes of electrodes. For applications where the vessels will be operating at low temperatures, smaller electrodes are used on multiple pass welds. This will produce smaller weld beads that build up the weld in relatively thin layers. The smaller weld beads give a stronger, tougher weld.

Industrial Piping

Shielded metal arc welding is widely used in the industrial piping industry which includes many types of pressure piping. The types of electrodes most often used are the E6010 and E7018 electrodes for welding low-carbon steel pipe. A common practice is the use of E6010 electrodes to weld in the root passes and the E7018 electrodes to weld in the fill and cover passes. Industrial piping is generally welded from the bottom to the top, except on small diameter pipe where it is done both ways. The reason that welding from bottom to top is most common is because slag is often trapped when welding in the opposite direction. For welding low-carbon steel pipe with a 70,000 psi (485 MPa) tensile strength, use E7010 and E7018 electrodes. Figure shows shielded metal arc welding with E7018 electrodes to weld structural supports.

Another example of this process is shown in where pulsed shielded metal arc welding is being used to cylindrical support beams. Shielded metal arc welding is often used for welding on other types of industrial piping. EXX15, EXX16, and EXX18 electrodes are used for welding chromium-molybdenum alloy pipe. When welding stainless steel pipe, gas tungsten arc welding (TIG) is often used to put in the root pass, and shielded metal arc welding is used to weld in the fill and cover passes. Medium and high-carbon steel pipe are also welded by this process. For these, smaller diameter electrodes are used than on low-carbon steels, in order to reduce the heat effect on the pipe.

E7018 electrode being used to weld structural supports.

SMAW cylindrical support beams.

Transmission Pipelines

The shielded metal arc welding process is by far the major process for welding on transmission or cross-country pipelines. Welding is done in the field, usually from the outside of the pipe, but whenever possible, welding should be done from both sides of the pipe. E6010 and E7018 electrodes are the types used for welding transmission pipelines. Several common procedures are in use for welding transmission pipelines. One of these is to put in the root pass with E6010 electrodes and put in the fill and cover passes with E7018 electrodes. Another is to use E7018's to weld in all passes, and a third is to weld in the root pass with the gas metal arc welding process and put in the rest of the passes with E7018 electrodes.

Nuclear Power Plants

The nuclear power industry employs this process for many applications. It is often used in the shop fabrication of low-carbon and low alloy steel heavy-walled pressure vessels and for welding longitudinal and circumferential weld seams. Shielded metal arc welding is the best method for welding nozzles and attachments to the vessels. A major application of this process is welding pressure piping for use in the nuclear power facilities. Nuclear power system pressure piping requires stronger quality control than normal pressure piping.

Structural welding with the SMAW process.

Structures

The construction industry is a major application for shielded metal arc welding. Most of the welding on buildings and bridges is done in the field at long distances from the power sources, which makes this process popular for these applications. Most types of covered electrodes are used in structural work because of the wide variety in the tensile strengths of the steels used. Figure shows structural welding with the shielded metal arc welding process. Figure shows a section of hinge being welded. In a Seabee is welding an Ibeam. Another example of the use of shielded metal arc welding in the construction industry is shown in where this process is being used to weld armor plating.

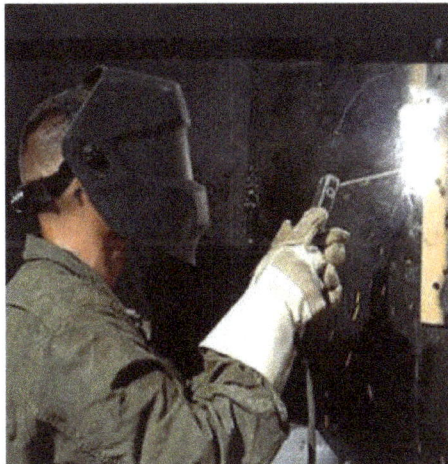

Hinge support welding.

I beam welding.

Ships

Shielded metal arc welding is still the major process used in shipbuilding. It is used for many different applications including welding in areas where the other processes cannot reach. Most types of low-carbon steel covered electrodes are used except the EXX12, EXX13, and EXX14 types. These three types of electrodes are not approved for use on the main structural members in the ship because of the relatively low ductility obtained from the weld deposits of these electrodes. The electrodes with large amounts of iron powder in their coatings are popular for many shipbuilding applications because of the high deposition rates obtained. These types of electrodes are especially used on the many fillet welds that are made in a ship structure. Backing tape is often used for backing the weld metal when one side welding is done.

Transportation

Another industry that widely uses this welding process is the railroad industry. It uses E60XX and E70XX electrodes to weld many parts of the underframe, cab, and engine of the locomotive. The underframe fabrication consists of mostly fillet welds. The frames and brackets for the diesel engines are also welded with these electrodes.

Railroad cars are commonly welded together by the shielded metal arc process. Underframes are often welded with E6020, E7016, and E7018 electrodes. The sills for the underframes are welded using E7024 electrodes because high deposition rates are desired for this application.

The automotive industry uses this process to a lesser extent. There, it is mainly used for welding low production components or on items where there are frequent model changes. This is because the fixtures and equipment for this process are less expensive.

Welding bars on a door.

Industrial Machinery

The frames of many types of heavy industrial machinery are welded together using this process. It is the major process used for welding piping associated with this machinery. Shielded metal arc welding is used for welding areas the other processes cannot reach. Figure shows a welder welding bars on a door.

Heavy Equipment

Another major application of this process is in the heavy equipment industry such as mining, agricultural, and earthmoving equipment. In these industries, shielded metal arc welding is used for welding structural steels, which are used for the frames, beams, and many other items in the assembly. Most types of covered electrodes are used depending on the type of steel being welded. Stainless steel and nonferrous metals are also used for some parts. Figure shows a Seabee welding a plate for a backhoe bucket.

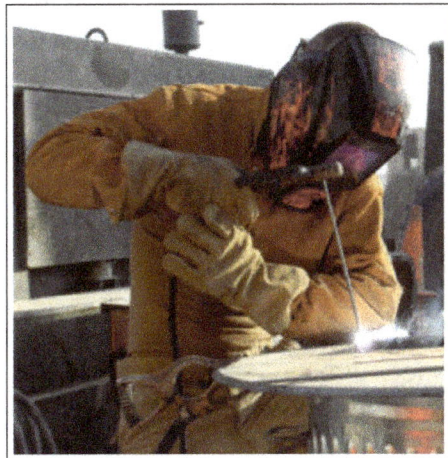

Welding a structural plate.

Maintenance and Repair

One industry where shielded metal arc welding is and will probably always remain the major welding process is the maintenance and repair industry. This is especially true in small shops and general plant maintenance, where relatively inexpensive equipment, portability, and versatility are important considerations. This process is the major one for repair welding on railroad engines and cars as well as cast iron engine blocks and heads on automobiles. This kind of welding commonly employs nickel electrodes for repairing cast iron parts. Resurfacing worn parts and putting a hard surface on parts (wearfacing) are two other applications. Special surfacing and build-up electrodes are used for these purposes.

Variations of the Process

Gravity welding, which is seldom used today, was an automatic variation of the shielded metal arc welding process. Gravity welding was popular because one operator could operate several gravity feeders at the same time, increasing the production rate. The welder installed the electrode in the feeder and the electrode fed as it burned off, which gave a high quality horizontal fillet weld. The welders usually used 28 in. (710 mm) long electrodes of the drag type (E6027, E7024, E7028).

These were used in diameters of 7/32 in. (5.6 mm) and in 1/4 in. (6.4 mm). This was possible in some shipbuilding work since the welds were often close together, which allowed the welding operator to quickly move from one holder to another to reload them, start them, and allow them to operate unattended.

Firecracker welding is a method of automatically making welds using a long electrode with an electrically nonconductive heavy coating. This method has been used very little in North America because of the popularity of semi-automatic processes. This method can be used for square groove butt welds and full fillet lap welds. To make a firecracker fillet weld, position the work flat. Place the welding electrode in the joint and place a retaining bar over it. Start the arc by shorting the end of the electrode to the work. The arc length depends on the thickness of the coating. As the arc travels along the electrode, the electrode melts and makes a deposit on the metal immediately underneath it. Once the arc is started, the process automatically proceeds to completion.

Another variation of shielded metal arc welding is the use of massive electrodes which have extremely large diameters and long lengths. These electrodes are so heavy that they require a manipulator to hold and feed them. Massive electrode welding is primarily used for repairing very large castings.

Flux Cored Arc Welding

Flux cored arc welding (FCAW) is an arc welding process in which the heat for welding is produced by an arc between a continuously fed tubular electrode wire and the work. Shielding is obtained by a flux contained within the tubular electrode wire or by the flux and an externally supplied shielding gas.

Flux cored arc welding is similar to gas metal arc welding in many ways, but the fluxcored wires used for this process give it different characteristics. Flux cored arc welding is widely used for welding ferrous metals and is particularly good for applications where high deposition rates are desirable. Also, at high welding currents, the arc is smooth and more manageable when compared to using large diameter gas metal arc welding electrodes with carbon dioxide. With FCAW, the arc and weld pool are clearly visible to the welder, and a slag coating is left on the surface of the weld bead, which must be removed. Since the filler metal transfers across the arc, some spatter is created and some smoke produced.

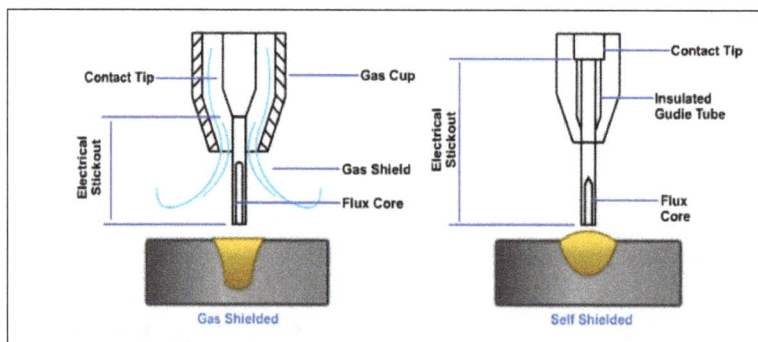

FCAW self shielded and external gas shielded electrodes.

As in GMAW, FCAW depends on a gas shield to protect the weld zone from detrimental atmospheric contamination. However, with FCAW, there are two primary ways this is accomplished:

- The gas is applied from an external source, in which case the electrode is referred to as a gas shielded flux-cored electrode.

- The gas is generated from the decomposition of gas-forming ingredients contained in the electrode's core. In this instance, the electrode is known as a self-shielding flux-cored electrode.

In addition to the gas shield, the flux-cored electrode produces a slag covering for further protection of the weld metal as it cools, which must be manually removed with a wire brush or chipping hammer.

The main advantage of the self-shielding method is that its operation is somewhat simplified because of the absence of external shielding equipment. Although self- shielding electrodes have been developed for welding low-alloy and stainless steels, they are most widely used on mild steels. The self-shielding method generally uses a long electrical stickout (distance between the contact tube and the end of the unmelted electrode, commonly from one to four inches). Electrical resistance is increased with the long extension, preheating the electrode before it is fed into the arc. This preheating enables the electrode to burn off at a faster rate and increases deposition. The preheating also decreases the heat available for melting the base metal, resulting in a more shallow penetration than the gas shielded process.

A major drawback of the self-shielded process is the metallurgical quality of the deposited weld metal. In addition to gaining its shielding ability from gas-forming ingredients in the core, the self-shielded electrode contains a high level of deoxidizing and denitrifying alloys, primarily aluminium, in its core. Although the aluminium performs well in neutralizing the effects of oxygen and nitrogen in the arc zone, its presence in the weld metal will reduce ductility and impact strength at low temperatures. For this reason, the self-shielding method is usually restricted to less critical applications.

The self-shielding electrodes are more suitable for welding in drafty locations than the gas-shielded types. Since the molten filler metal is on the outside of the flux, the gases formed by the decomposing flux are not totally relied upon to shield the arc from the atmosphere. To compensate, the deoxidizing and denitrifying elements in the flux further help to neutralize the effects of nitrogen and oxygen present in the weld zone.

The gas-shielded flux-cored electrode has a major advantage over the self-shielded flux-cored electrode, which is, the protective envelope formed by the auxiliary gas shield around the molten puddle. This envelope effectively excludes the atmosphere without the need for core ingredients, such as aluminium. Because of this more thorough shielding, the weld metallurgy is cleaner, which makes this process suitable for welding not only mild steels, but also low-alloy steels in a wide range of strength and impact levels.

The gas-shielded method uses a shorter electrical stickout than the self-shielded process. Extensions from 1/2" to 3/4" are common on all diameters, and 3/4" to 1-1/2" on larger diameters. Higher welding currents are also used with this process, enabling high deposition rates. The auxiliary shielding helps to reduce the arc energy into a columnar pattern. The combination of high currents and the action of the shielding gas contributes to the deep penetration inherent with this process. Both spray and globular transfer are utilized with the gas-shielded process.

Methods of Application

Although flux cored arc welding may be applied semiautomatically, by machine, or automatically, the process is usually applied semiautomatically. In semiautomatic welding, the wire feeder feeds the electrode wire and the power source maintains the arc length. The welder manipulates the welding gun and adjusts the welding parameters. FCAW is also used in machine welding where, in addition to feeding the wire and maintaining the arc length, the machinery also provides the joint travel. The welding operator continuously monitors the welding and makes adjustments in the welding parameters. Automatic welding is used in high production applications. In automatic welding, the welding operator only starts the operation.

Advantages and Limitations

Flux cored arc welding has many advantages for a wide variety of applications. It often competes with shielded metal arc welding, gas metal arc welding, and submerged arc welding (SAW) for many applications. Some of the advantages of this process are:

- It has a high deposition rate and faster travel speeds.

- Using small diameter electrode wires, welding can be done in all positions.

- Some flux-cored wires do not need an external supply of shielding gas, which simplifies the equipment.

- The electrode wire is fed continuously so there is very little time spent on changing electrodes.

- Deposits a higher percentage of the filler metal when compared to shielded metal arc welding.

- Obtains better penetration than shielded metal arc welding.

Principles of Operation

Flux cored arc welding uses the heat of an electric arc between a consumable, tubular electrode and the part to be welded. Electric current passing through an ionized gas produces an electric arc. The gas atoms and molecules are broken up and ionized by losing electrons and leaving a positive charge. The positive gas ions then flow from the positive pole to the negative pole and the electrons flow from the negative pole to the positive pole. The electrons carry about 95% of the heat and the rest is carried by the positive ions. The heat of the arc melts the electrode and the surface of the base metal.

One of two methods shields the molten weld metal, heated weld zone, and electrode. The first method is by the decomposition of the flux core of the electrode. The second method is by a combination of an externally supplied shielding gas and the decomposition of the flux core of the electrode wire. The flux core has essentially the same purpose as the coating on an electrode for shielded metal arc welding. The molten electrode filler metal transfers across the arc and into the molten weld puddle, and a slag forms on top of the weld bead that can be removed after welding.

The arc is struck by starting the wire feed which causes the electrode wire to touch the workpiece and initiate the arc. Arc travel is usually not started until a weld puddle is formed. The welding gun then moves along the weld joint manually or mechanically so that the edges of the weld joint are joined. The weld metal then solidifies behind the arc, completing the welding process. A large amount of flux is contained in the core of a selfshielding wire as compared to a gas-shielded wire. This is needed to provide adequate shielding and because of this, a thicker slag coating is formed. In these wires, deoxidizing and denitrifying elements are needed in the filler metal and flux core because some nitrogen is introduced from the atmosphere.

Arc Systems

The FCAW process may be operated on both constant voltage and constant current power sources. A welding power source can be classified by its volt-ampere characteristics as a constant voltage (also called constant potential) or constant current (also called variable voltage) type, although there are some machines that can produce both characteristics. Constant voltage power sources are preferred for a majority of FCAW applications.

In the constant voltage arc system, the voltage delivered to the arc is maintained at a relatively constant level that gives a flat or nearly flat volt-ampere curve, as shown in This type of power source is widely used for the processes that require a continuously fed wire electrode. In this system, the arc length is controlled by setting the voltage level on the power source and the welding current is controlled by setting the wire feed speed.

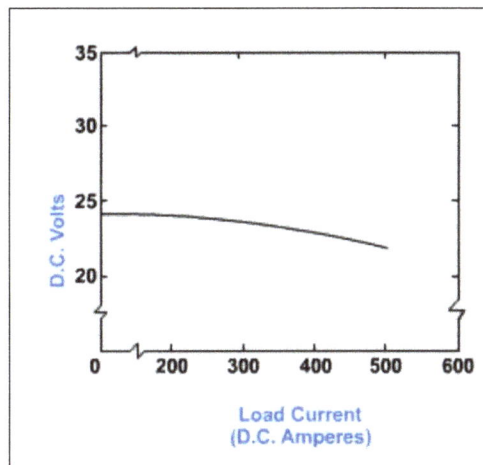

Constant voltage system volt-ampere curve.

As figure shows a slight change in the arc length (voltage level) will produce a large change in the welding current.

Most power sources have a fixed slope built in for a certain type of flux cored arc welding. Some constant voltage welding machines are equipped with a slope control used to change the slope of the voltampere curve.

The figure shows different slopes obtained from one power source. The slope has the effect of limiting the amount of shortcircuiting current the power supply can deliver. This is the current available from the power source on the short-circuit between the electrode wire and the work. This is not as important in FCAW as it was in GMAW because short-circuiting metal transfer is not encountered except with alloy cored, low flux content wires.

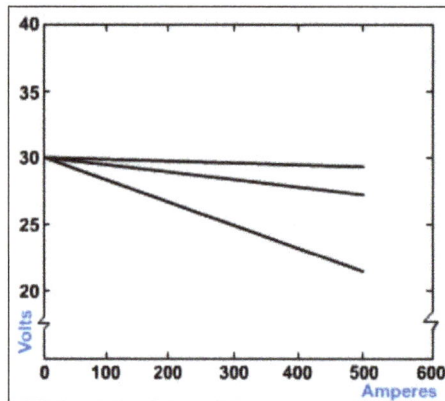

Different slopes from a constant voltage motor generator power source.

A slope control is not required, but may be desirable, when welding with small diameter, alloy cored, low flux content electrodes at low current levels. The shortcircuit current determines the amount of pinch force available on the electrode. The pinch forces cause the molten electrode droplet to separate from the solid electrode. The flatter the slope of the volt-ampere curve, the higher the short-circuit and the pinch force. The steeper the slope, the lower the short-circuit and pinch force. The pinch force is important with these electrodes because it affects the way the droplet detaches from the tip of the electrode wire. When a high short-circuit and a flat slope cause pinch force, excessive spatter is created. When a very low short-circuit current and pinch force are caused by a steep slope, the electrode wire tends to freeze in the weld puddle or pile up on the work piece. When the proper amount of short-circuit current is used, it creates very little spatter.

The inductance of the power supply also has an effect on the arc stability. When the load on the power supply changes, the current takes time to find its new level. The rate of current change is determined by the inductance of the power supply. Increasing the inductance will reduce the rate of current rise. The rate of the welding current rise increases with the current that is also affected by the inductance in the circuit. Increased arc time or inductance produces a flatter and smoother weld bead as well as a more fluid weld puddle. Too much inductance will cause more difficult arc starting.

The constant current arc system provides a nearly constant welding current to the arc, which gives a drooping volt-ampere characteristic, as shown in.This arc system is used with the SMAW and GTAW processes. A dial on the machine sets the welding current and the welding voltage is controlled by the arc length held by the welder.

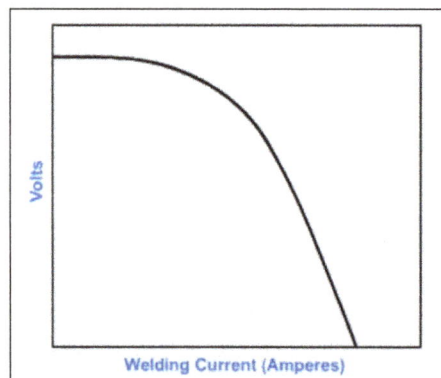

Volt-ampere curve for a constant current arc system.

This system is necessary for manual welding because the welder cannot hold a constant arc length, which causes only small variations in the welding current. When flux cored arc welding is done with a constant current system, a special voltagesensing wire feeder is used to maintain a constant arc length.

For any power source, the voltage drop across the welding arc is directly dependent on the arc length. An increase in the arc length results in a corresponding increase in the arc voltage and a decrease in the arc length results in a corresponding decrease in the arc voltage.

Another important relationship exists between the welding current and the melt off-rate of the electrode. With low current, the electrode melts off slower and the metal is deposited slower. This relationship between welding current and wire feed speed is definite, based on the wire size, shielding gas type and type of electrode. A faster wire feed speed will give a higher welding current.

In the constant voltage system, instead of regulating the wire to maintain a constant arc length, the wire is fed into the arc at a fixed speed and the power source is designed to melt off the wire at the same speed. The self-regulating characteristic of a constant voltage power source comes about by the ability of this type of power source to adjust its welding current in order to maintain a fixed voltage across the arc.

With the constant current arc system, the welder changes the wire feed speed as the gun is moved toward or away from the weld puddle. Since the welding current remains the same, the burn-off rate of the wire is unable to compensate for the variations in the wire feed speed, which allows stubbing or burning back of the wire into the contact tip to occur. To lessen this problem, a special voltage-sensing wire feeder is used, which regulates the wire feed speed to maintain a constant voltage across the arc.

The constant voltage system is preferred for most applications, particularly for small diameter wire. With smaller diameter electrodes, the voltage-sensing system is often unable to react fast enough to feed at the required burn-off rate, resulting in a higher instance of burnback into the contact tip of the gun.

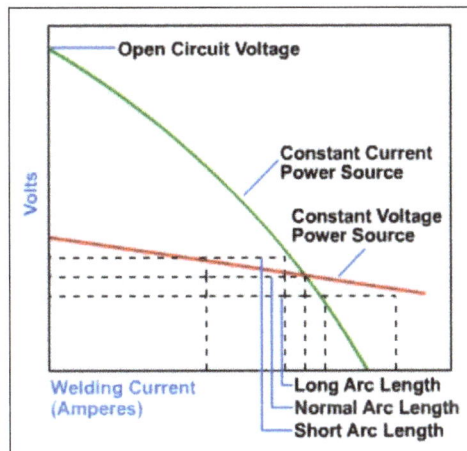

Volt-ampere curves.

The figure shows a comparison of the voltampere curves for the two arc systems. This shows that for these particular curves, when a normal arc length is used, the current and voltage levels are the same for both the constant current and constant voltage systems. For a long arc length, there is a slight drop in the welding current for the constant current machine and large drop in the current for a constant voltage machine. For constant voltage power sources, the volt-ampere curve shows that when the arc length shortens slightly, a large increase in welding current occurs. This results in an increased burn-off rate, which brings the arc length back to the desired level. Under this system, changes in the wire feed speed, caused by the welder, are compensated for electrically by the power source.

Metal Transfer

Metal transfer, from consumable electrodes across an arc, has been classified into three general modes of transfer: spray transfer, globular transfer, and short-circuiting transfer. The metal transfer of most flux-cored electrodes resembles a fine globular transfer. Only the alloy-cored, low flux content wires can produce a short-circuiting metal transfer similar to GMAW.

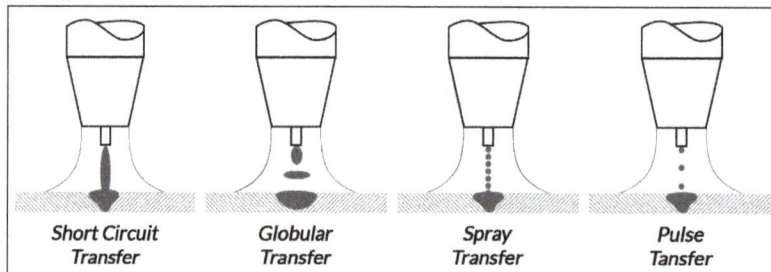

| Short Circuit Transfer | Globular Transfer | Spray Transfer | Pulse Tansfer |

On flux-cored electrodes, the molten droplets build up around the periphery or outer metal sheath of the electrode. By contrast, the droplets on solid wires tend to form across the entire cross section at the end of the wire. A droplet forms on the cored wire, is transferred, and then a droplet is formed at another location on the metal sheath. The core material appears to transfer independently to the surface of the weld puddle. The metal transfer in flux cored arc welding.

At low currents, the droplets tend to be larger than at higher current levels. If the welding current using a 3/32 in. (2.4 mm) electrode wire is increased from 350 to 550 amps, the metal transfer characteristics will change. Transfer is much more frequent and the droplets become smaller as the current is increased. At 550 amperes, some of the metal may transfer by the spray mode, although the globular mode prevails. There is no indication that higher currents cause a transition to a spray mode of transfer, unless an argon-oxygen shielding gas mixture is used.

The larger droplets at the lower currents cause a certain amount of "splashing action" when they enter the weld puddle. This action decreases with the smaller droplet size.

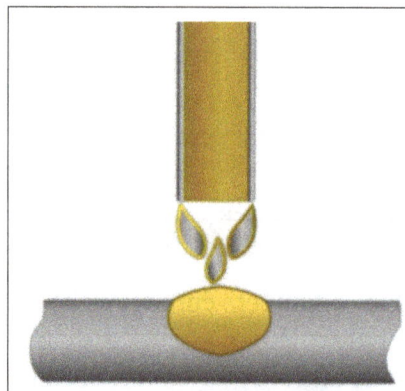

Metal transfer in FCAW.

Equipment for Welding

The equipment used for FCAW is very similar to that used for GMAW. The basic arc welding equipment consists of a power source, controls, wire feeder, welding gun, and welding cables. A major difference between the gas-shielded electrodes and self - shielded electrodes is that the gas shielded wires also require a gas shielding system. This may also have an effect on the type of welding gun used. Fume extractors are often used with this process. For machine and automatic welding, several items, such as seam followers and motion devices, are added to the basic equipment. A figure of the equipment for semiautomatic FCAW is shown in figure.

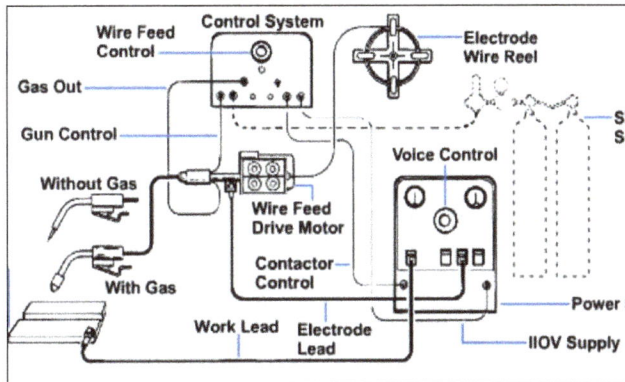

Equipment for flux cored arc welding.

Power Sources

The power source (welding machine) provides the electric power of the proper voltage and amperage to maintain a welding arc. Most power sources operate on 230 or 460 volt input power, but machines that operate on 200 or 575 volt input are available as options. Power sources may operate on either single-phase or three-phase input with a frequency of 50 to 60 Hz.

Power Source Duty Cycle

Duty cycle is defined as the ratio of arc time to total time. Most power sources used for FCAW have a duty cycle of 100%, which indicates that they can be used to weld continuously. However, some machines have a duty cycle of 60%. For a welding machine, a 10 minute time period is used. Thus, for a 60% duty cycle machine, the welding load would be applied continuously for 6 minutes and would be off for 4 minutes. Most industrial type, constant current machines are rated at 60% duty cycle. The formula for determining the duty cycle of a welding machine for a given load current is:

$$\%\text{Duty Cycle} = \frac{(\text{Rated Current})^2}{(\text{Load Current})^2} \times \text{Rated Duty Cycle}$$

For example, if a welding machine is rated at a 60% duty cycle at 300 amperes, the duty cycle of the machine when operated at 350 amperes would be:

$$\%\text{Duty Cycle} = \frac{(300)^2}{(350)^2} \times 60 = 44\%$$

In general, these lower duty cycle machines are the constant current type, which are used in plants where the same machines are also used for SMAW and gas tungsten arc welding. Some of the smaller constant voltage welding machines have a 60% duty cycle.

Types of Current

FCAW uses direct current, which can be connected in one of two ways: electrode positive (reverse polarity) or electrode negative (straight polarity). The electrically charged particles flow between the tip of the electrode and the work as shown in figure.

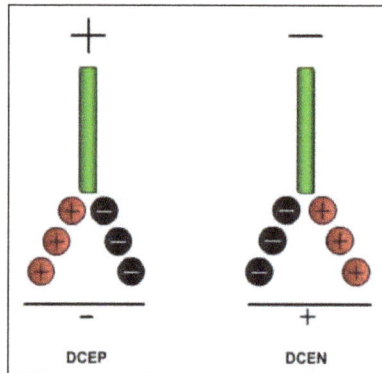

Particle Flow for DCEP and DCEN

Flux-cored electrode wires are designed to operate on either DCEP or DCEN. The wires designed for use with an external gas shielding system are generally designed for use with DCEP, while some self-shielding flux-cored wires are used with DCEP and others are used with DCEN. Electrode positive current gives better penetration into the weld joint. Electrode negative current gives lighter penetration, and is used for welding thinner metal or where there is poor fit-up. The weld created by DCEN is wider and shallower than the weld produced by DCEP.

Types of Power Sources

The power sources generally recommended for flux cored arc welding are direct current constant voltage types. Both rotating (generator) and static (single- or three-phase transformer-rectifiers) are used. Any of these types of machines are available to produce constant current or constant voltage output, or both. The same power sources used with GMAW are used with FCAW, but FCAW generally uses higher welding currents, which sometimes requires a larger power source. It is important to use a power source capable of producing the maximum current level required for an application.

Generator and Alternator Welding Machines

Generator welding machines used for this process can be powered by an electric motor for shop use, or an internal combustion engine for field applications. Gasoline or diesel engine-driven welding machines have either liquid or air-cooled engines and many of them provide auxiliary power for emergency lighting, power tools, etc. Many of the engine-driven generators used for FCAW in the field are combination constant current-constant voltage types. These types are popular for applications where both SMAW and FCAW can be accomplished using the same power source. Figure shows an engine-driven generator machine used for flux cored arc welding. The motor-driven generator welding machines are gradually being replaced by transformer-rectifier welding machines. Motor-driven generators produce a very stable arc, but they are noisier, more expensive, consume more power and require more maintenance than transformer-rectifier machines. They can, however, function without being sourced by an electrical power supply and, in fact, can produce the auxiliary electricity during power outages.

Gas powered welder/generator.

An alternator welding machine is an electric generator made to produce AC power. This power source has a rotating assembly. These machines are also called rotating or revolving field machines.

Transformer Welding Machines

Transformer-rectifiers are the most widely used welding machines for FCAW. Adding a rectifier to a basic transformer circuit is a method of supplying direct current to the arc without using a rotating generator. A rectifier is an electrical device which changes alternating current into direct current. These machines are more efficient electrically than motor-generator welding machines and they provide quieter operation. There are two basic types of transformer-rectifier welding machines: those that operate on singlephase input power and those that operate on three-phase input power.

The single-phase transformer-rectifier machines provide DC current to the arc and a constant current volt-ampere characteristic, but are not as popular as three-phase transformer-rectifier welding machines for FCAW. When using a constant current power source, a special variable speed or voltage-sensing wire feeder must be used to maintain a uniform current level. A limitation of the single-phase system is that the power required by the single-phase input power may create an unbalance of the power supply lines which is objectionable to most power companies. These machines normally have a duty cycle of 60%.

The most widely used type of power source for this process is the three-phase transformer-rectifier. These machines produce DC current for the arc, and for FCAW, most have a constant voltage voltampere characteristic. When using these constant voltage machines, a constantspeed wire feeder is used. This type of wire feeder maintains a constant wire feed speed with slight changes in welding current. The three-phase input power gives these machines a more stable arc than singlephase input power and avoids the line unbalance that occurs with the single-phase machines.

Many of these machines also use solid state controls for the welding. A 650 amp solid state controlled power source is shown in. This machine will produce the flattest volt-ampere curve of the different constant voltage power sources. Most threephase transformer-rectifier power sources are rated at a 100% duty cycle.

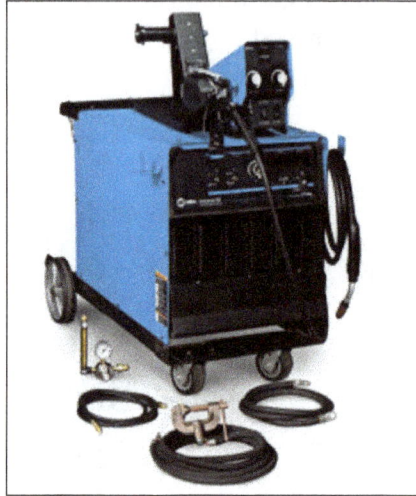

Three-phase, 650 amp solid state power source.

Controls

Programmable control unit.

The controls for this process are located on the front of the welding machine, on the welding gun, and on the wire feeder or a control box.

The welding machine controls for a constant voltage machine include an on-off switch, a voltage control, and often a switch to select the polarity of direct current. The voltage control can be a single knob, or it can have a tap switch for setting the voltage range and a fine-voltage control knob.

Other controls are sometimes present, such as a switch for selecting constant current (CC) or constant voltage (CV) output on combination machines, or a switch for a remote control. On constant current welding machines, there is an on-off switch, a current level control knob, and sometimes a knob or switch for selecting the polarity of direct current.

The trigger or switch on the welding gun is a remote control used by the welder in semiautomatic welding to stop and start the welding current, wire feed, and shielding gas flow. For semiautomatic welding, a wire feed speed control is normally part of, or close by, the wire feeder assembly. The wire feed speed sets the welding current level on a constant voltage machine. For machine or automatic welding, a separate control box is often used to control the wire feed speed. A control box for semiautomatic or automatic welding is shown in. There may also be switches to turn the control on and off on the wire feeder control box, and gradually feed the wire up and down.

Other controls for this process are used for special applications, especially when a programmable power source is used. An example is a timer for spot welding. Controls that produce a digital readout are popular because it is easier for concise control.

Wire Feeders

The wire feed motor provides the power for driving the electrode through the cable and gun to the work. There are several different wire feeding systems available. The selection of the best type of system depends on the application. Most FCAW wire feed systems are the constant speed type, which are used with constant voltage power sources. This means the wire feed speed is set before welding. The wire feed speed controls the amount of welding current. Variable speed or voltage-sensing wire feeders are used with constant current power sources. With a variable speed wire feeder, a voltage-sensing circuit maintains the desired arc length by varying the wire feed speed. Variations in the arc length increase or decrease the wire feed speed.

A wire feeder consists of an electrical motor connected to a gear box containing drive rolls. The gear box and wire feed motor shown in have four feed rolls in the gear box. While many systems have only two, in a four-roll system, the lower two rolls drive the wire.

Because of their structure, flux-cored wires can be easily flattened. The type of drive roll used is based on the size of the tubular wire being fed. The three basic types of drive rolls are the "U" groove, "V" knurled, and "U" cogged, as shown in the figure. "U" groove drive rolls are only used on small diameter wires. These can be used because small diameter tubular wires are less easily flattened. "V" knurled drive rolls are most commonly used for wire sizes 1/16 in. (1.6 mm) and greater. These drive rolls are lightly knurled to prevent slipping of the wire. The "U" cogged drive rolls are used for large diameter flux-cored wires. A groove is cut into both rolls. Different gear ratios are used, depending on the wire feed speed required. The wire feed speeds that can be obtained from different gear ratios.

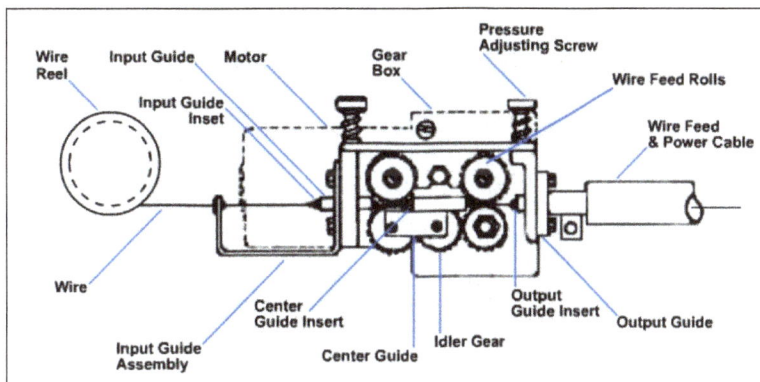

Wire feed assembly.

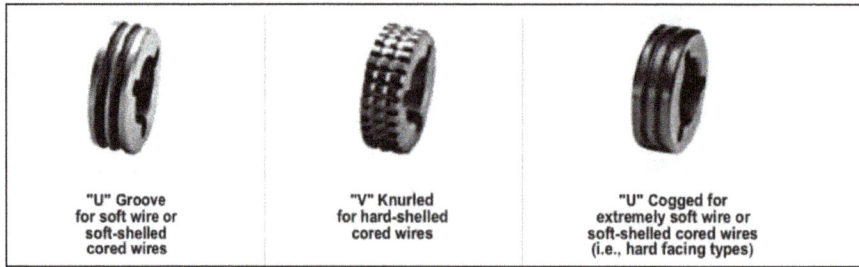

| "U" Groove for soft wire or soft-shelled cored wires | "V" Knurled for hard-shelled cored wires | "U" Cogged for extremely soft wire or soft-shelled cored wires (i.e., hard facing types) |

Drive roll types and applications.

Wire feed speeds obtained from different gear ratios:

		Wire Feed Speed
Gear Ratio	In/min	(mm/s)
15:1	500-2000	212-846
37.5:1	60-1000	25-423
46:1	50-825	21-349
75:1	30-500	13-212
90:1	25-400	11-169
150:1	15-250	6-106
300:1	8-125	3-53
600:1	4-63	2-27
1200:1	2-30	1-13

Wire feed systems may be the pull, push, or push-pull type, depending on the method of application and the distance between the welding gun and the coil or spool of wire. Pull type wire feeders have the drive rolls attached to the welding gun. Most machine and automatic welding stations use this type of system, but pull type wire feeders are rarely used in semiautomatic welding. Pull wire feeders have the advantage for welding small diameter aluminium and soft nonferrous metals with GMAW because it reduces wire feeding problems, but, since most flux-cored wires are steel, this is not an advantage for FCAW.

Semi-automatic, solid state control wire feeder.

The push type system with the drive rolls mounted near the coil or spool of wire is the most commonly used wire feed method for semiautomatic welding. The wire is pulled from the coil or spool and then pushed into a flexible conduit and through the gun. The relatively large diameter wires used in FCAW are well suited to this type of system. The length of the conduit can be up to about 12 feet (3.7 m). Another advantage of this push type system is that the wire feed mechanism is not attached to the gun, which reduces the weight and makes the gun easier to handle.

Some wire feed systems contain a two-gun, two wire feeder arrangement connected to a single control box, which is connected to a single power source. Both wire feeders may be set up, and there is a switch on the control to automatically select which of the two systems will be used.

One advantage to this system is that the second wire feeder and gun can provide backup in case of breakdown, gun maintenance, or electrode change. Another advantage is that two different electrodes for different applications can be set up. For example, a GMAW electrode and gun can be set up on one schedule for welding a root pass, and a second schedule can be set up with a flux-cored wire to weld the rest of the joint with FCAW's faster deposition. This eliminates the need for two power sources or the need to change the electrode wire and gun. The liner is made of flexible metal and is available in sizes compatible with the electrode size. The liner guides the electrode wire from the wire feeder drive rolls through the cable assembly and prevents interruptions in the travel.

Heavy-duty welding guns are normally used because of the large size electrode wires typically used and the corresponding high welding current levels required. Because of the intense heat created by this process, heat shields are attached to the gun in front of the trigger to protect the welder's hand.

Both air-cooled and water-cooled guns are used for FCAW. Air-cooled guns are cooled primarily by the surrounding air, but when a shielding gas is used, this will have an additional cooling effect.

A water-cooled gun is similar to an air-cooled gun, except that ducts to permit the water to circulate around the contact tube and nozzle have been added. Water-cooled guns permit more efficient cooling of the gun. Figure shows a 500-ampere watercooled gun. Water-cooled guns are preferred for many applications using 500 amperes and recommended for use with welding currents greater than 600 amperes. Welding guns are rated at the maximum current capacity for continuous operation.

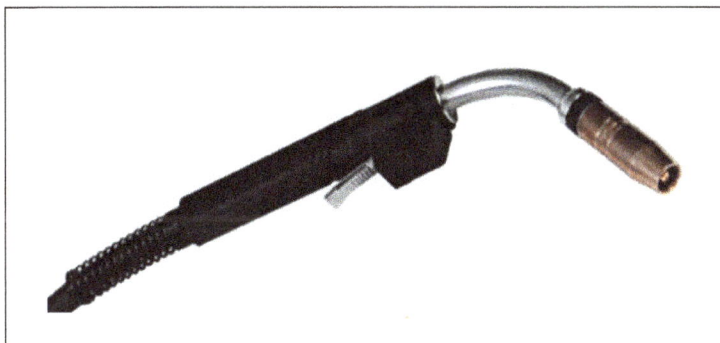
Water-cooled gun.

Air-cooled guns are lighter and easier to manipulate. The figure shows a 350 ampere air-cooled welding gun.

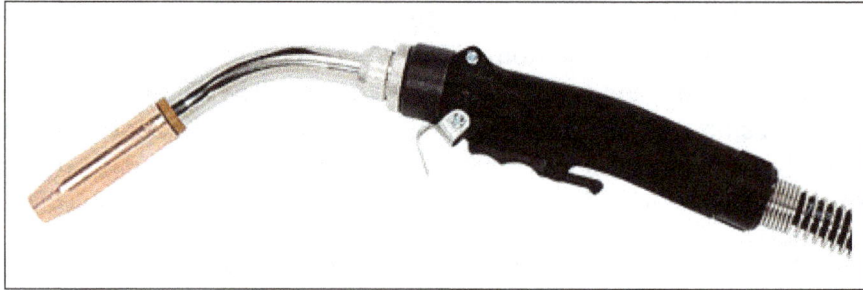

Air-cooled gun.

Some self-shielded electrode wires require a specific minimum electrode extension to develop proper shielding, so welding guns for these electrodes have guide tubes with an insulated extension guide. This guide supports the electrode and insures a minimum electrode extension, as shown in.

Machine Welding Guns

Machine and automatic welding guns use the same basic design principles and features as the semiautomatic welding guns. These guns often have very high current-carrying capacities and may also be air cooled or watercooled. Large diameter wires up to 1/8 in. (3.2 mm) are commonly used with high amperages. Machine welding guns must be heavy duty because of the high amperages and duty cycles required, and the welding gun is mounted directly below the wire feeder. Figure shows a machine welding head for FCAW.

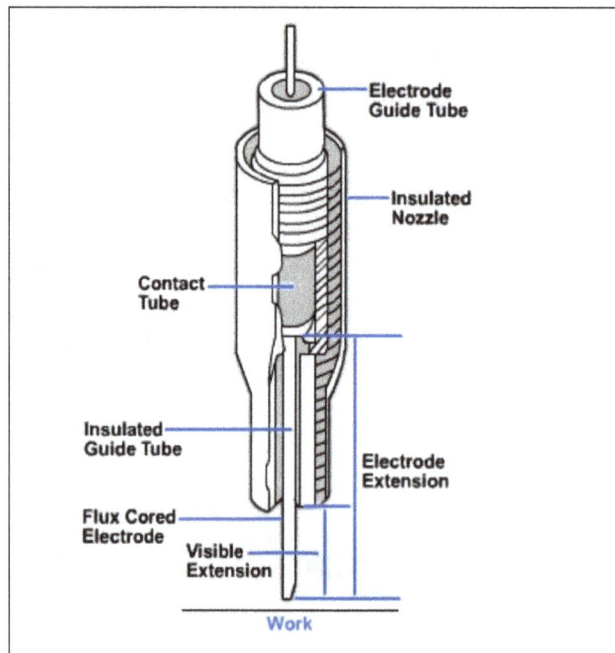

Insulated extension guide.

If a gas-shielded wire is to be used, the gas can be supplied by a nozzle that is concentric around the electrode or by a side delivery tube, as is shown in. The side shielding permits the welding gun to be used in deep, narrow grooves and reduces spatter buildup problems in the nozzle. Side shielding is only recommended for welding using carbon dioxide. A concentric nozzle is preferred when using argon-carbon dioxide and argon-oxygen mixtures, and a concentric nozzle provides better shielding and is sometimes recommended for CO_2 at high current levels when a large weld puddle exists.

Weld and Joint Design

Like other welding processes, the weld joint designs used in FCAW are determined by the design of the weldment, metallurgical considerations, and codes or specifications. Another factor to consider is the method of FCAW to be used. A properly selected joint design should allow the highest quality weld to be made at the lowest possible cost. A weld joint consists of a specific weld being made in a specific joint. A joint is defined as the junction of members which are to be, or have been, joined. Figure shows the five basic joint classifications.

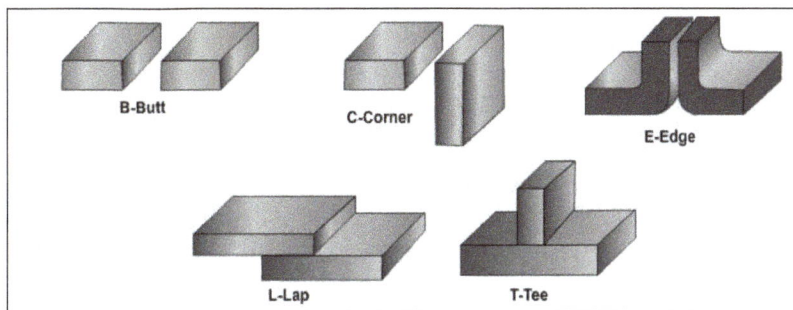

Types of joints.

Each of the different types of joints can be joined by many different types of welds. The figure shows the most common types of welds made.

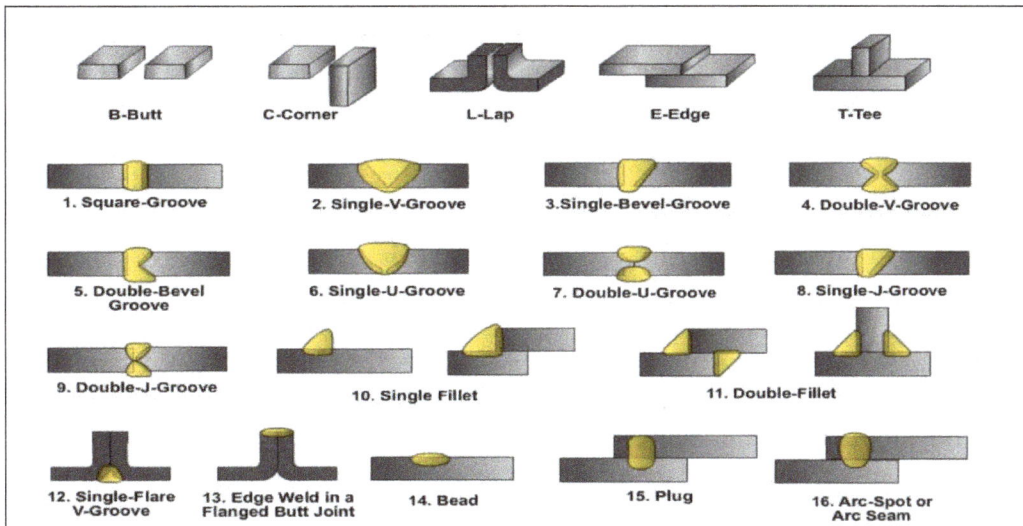

Types of welds.

The type of weld made is governed by the joint configuration. Each of the different types of welds has its own specific advantages. The nomenclature used for the various parts of groove and fillet welds is given in. There are several factors that influence the joint design to be used:

- Process Method.

- Strength Required.

- Welding Position.

- Joint Accessibility.

- Metal Thickness.

- Type of Metal.

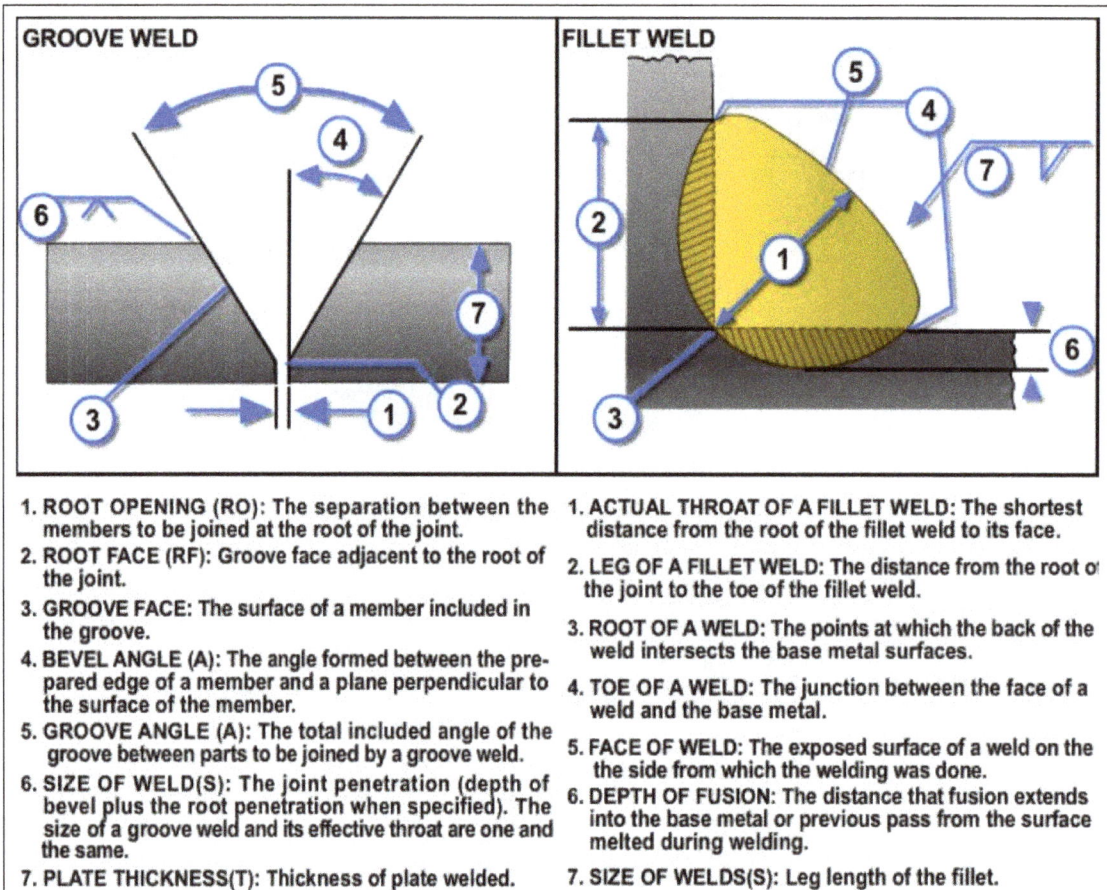

GROOVE WELD

FILLET WELD

1. **ROOT OPENING (RO):** The separation between the members to be joined at the root of the joint.

2. **ROOT FACE (RF):** Groove face adjacent to the root of the joint.

3. **GROOVE FACE:** The surface of a member included in the groove.

4. **BEVEL ANGLE (A):** The angle formed between the prepared edge of a member and a plane perpendicular to the surface of the member.

5. **GROOVE ANGLE (A):** The total included angle of the groove between parts to be joined by a groove weld.

6. **SIZE OF WELD(S):** The joint penetration (depth of bevel plus the root penetration when specified). The size of a groove weld and its effective throat are one and the same.

7. **PLATE THICKNESS(T):** Thickness of plate welded.

1. **ACTUAL THROAT OF A FILLET WELD:** The shortest distance from the root of the fillet weld to its face.

2. **LEG OF A FILLET WELD:** The distance from the root of the joint to the toe of the fillet weld.

3. **ROOT OF A WELD:** The points at which the back of the weld intersects the base metal surfaces.

4. **TOE OF A WELD:** The junction between the face of a weld and the base metal.

5. **FACE OF WELD:** The exposed surface of a weld on the the side from which the welding was done.

6. **DEPTH OF FUSION:** The distance that fusion extends into the base metal or previous pass from the surface melted during welding.

7. **SIZE OF WELDS(S):** Leg length of the fillet.

Weld nomenclature.

The edge and joint preparation are important because they affect both the quality and cost of welding. The cost items to be considered are the amount of filler metal required, the method of joint preparation, the amount of labor required, and the quality level required. Joints that are more difficult to weld will often have more repair work necessary than those that are easier to weld. This can lead to significant increases in cost, since repair welding sometimes requires more time and expense than the original weld. All of the five basic joint types are applicable to FCAW, although the butt and Tjoints are the most widely used. Lap joints have the advantage of not requiring much preparation other than squaring off the edges and making sure the members are in close contact. Edge joints are widely used on thin metal. Corner joints generally use similar edge preparations to those used on T-joints.

Many of the joint designs used for FCAW are similar to those used in GMAW or SMAW. FCAW has some characteristics that may affect the joint design. The joint should be designed so the welder has good access to the joint and is properly able to manipulate the electrode. Joints must be located so an adequate distance between the joint and nozzle of the welding gun is created. The proper distance will vary depending on the type of flux-cored electrode being used.

Process Method

The joint design as well as the welding procedure will vary, depending on whether the welding is done using gas-shielded or self-shielded electrodes. Both methods of FCAW achieve deeper penetration than SMAW. This permits the use of narrower grooves with smaller groove angles, larger root faces, and narrower root openings. Differences also exist between the two FCAW methods because of the deeper penetration that is produced by the gas-shielded electrode wires. Figure shows a comparison of a flat position, V-groove weld on a backing strip for each of the two methods. The joint design for the self-shielding wire requires a larger root opening to allow better access to the root of the joint. The joint design for the gas-shielded wire does not need such a wide root opening because complete penetration is easier to obtain. This weld would be less expensive to make using the gas-shielded electrode because less filler metal is required. This difference in joint design usually only applies when a backing strip is used.

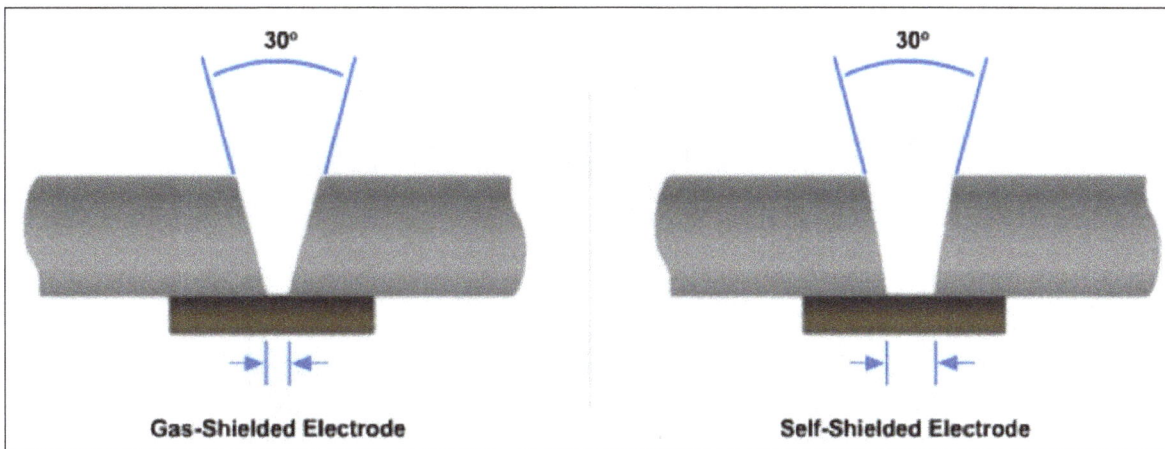

30°   **30°**

Gas-Shielded Electrode   **Self-Shielded Electrode**

Comparison between gas-shielded and self-shielded wire joint designs for the flat position.

For joints not requiring a backing strip, gas-shielded and self-shielded wires use the same joint designs.

Type of Metal

The FCAW process is used to weld steel, some stainless steels, and some nickels. The influence of the type of metal on the joint design is based primarily on the physical properties of the metal to be welded and whether or not the metal has an oxide coating. For example, stainless steels have a lower thermal conductivity than carbon steels. This causes the heat from welding to remain in the weld zone longer, which enables slightly greater thicknesses of stainless steels to be welded using a square groove joint design. Stainless steels also have an oxide coating that tends to reduce the depth of fusion of the weld. Consequently, stainless steels normally use larger groove angles and root openings than carbon steels. This allows the welder to direct the arc on the base metal surfaces to obtain complete fusion.

Gas Tungsten Arc Welding

Gas tungsten arc welding (GTAW), also known as tungsten inert gas (TIG) welding, is an arc welding process that uses a nonconsumable tungsten electrode to produce the weld. The weld area is protected from atmospheric contamination by a shielding gas (usually an inert gas such as argon), and a filler metal is normally used, though some welds, known as autogenous welds, do not require it. A constant-current welding power supply produces energy which is conducted across the arc through a column of highly ionized gas and metal vapors known as a plasma.

GTAW is most commonly used to weld thin sections of stainless steel and light metals such as aluminium, magnesium, and copper alloys. The process grants the operator greater control over the weld than competing procedures such as shielded metal arc welding and gas metal arc welding, allowing for stronger, higher quality welds. However, GTAW is comparatively more complex and difficult to master, and furthermore, it is significantly slower than most other welding techniques. A related process, plasma arc welding, uses a slightly different welding torch to create a more focused welding arc and as a result is often automated.

After the discovery of the electric arc in 1800 by Humphry Davy, arc welding developed slowly. C. L. Coffin had the idea of welding in an inert gas atmosphere in 1890, but even in the early 1900s, welding non-ferrous materials like aluminium and magnesium remained difficult, because these metals reacted rapidly with the air, resulting in porous and dross-filled welds.

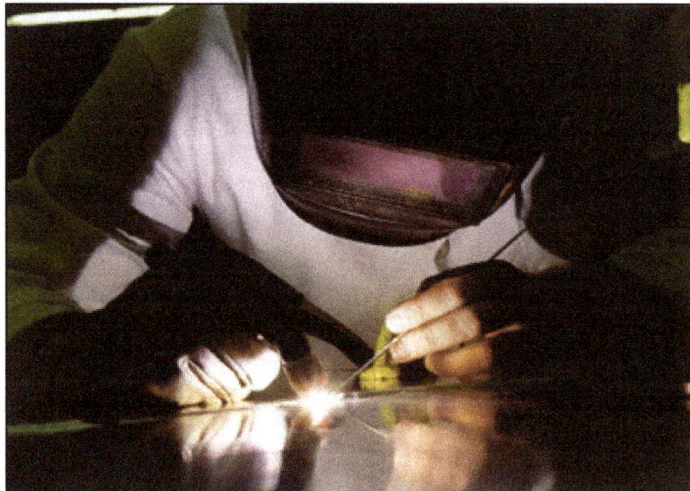

Gas Tungsten Arc Welding.

Processes using flux covered electrodes did not satisfactorily protect the weld area from contamination. To solve the problem, bottled inert gases were used in the beginning of the 1930s. A few years later, a direct current, gas-shielded welding process emerged in the aircraft industry for welding magnesium.

This process was perfected in 1941, and became known as heliarc or tungsten inert gas welding, because it utilized a tungsten electrode and helium as a shielding gas. Initially, the electrode overheated quickly, and in spite of tungsten's high melting temperature, particles of tungsten were transferred to the weld. To address this problem, the polarity of the electrode was changed from positive to negative, but this made it unsuitable for welding many non-ferrous materials. Finally, the development of alternating current made it possible to stabilize the arc and produce high quality aluminium and magnesium welds.

Developments continued during the following decades. Linde Air Products developed water-cooled torches that helped to prevent overheating when welding with high currents. Additionally, during the 1950s, as the process continued to gain popularity, some users turned to carbon dioxide as an alternative to the more expensive welding atmospheres consisting of argon and helium. However, this proved unacceptable for welding aluminium and magnesium because it reduced weld quality, and as a result, it is rarely used with GTAW today.

In 1953, a new process based on GTAW was developed, called plasma arc welding. It affords greater control and improves weld quality by using a nozzle to focus the electric arc, but is largely limited to automated systems, whereas GTAW remains primarily a manual, hand-held method. Development within the GTAW process has continued as well, and today a number of variations exist. Among the most popular are the pulsed-current, manual programmed, hot-wire, dabber, and increased penetration GTAW methods.

Operation

Manual gas tungsten arc welding is often considered the most difficult of all the welding processes commonly used in industry. Because the welder must maintain a short arc length, great care and skill are required to prevent contact between the electrode and the workpiece. Unlike other welding processes, GTAW normally requires two hands, since most applications require that the welder manually feed a filler metal into the weld area with one hand while manipulating the welding torch in the other. However, some welds combining thin materials (known as autogenous or fusion welds) can be accomplished without filler metal; most notably edge, corner and butt joints.

To strike the welding arc, a high frequency generator provides a path for the welding current through the shielding gas, allowing the arc to be struck when the separation between the electrode and the workpiece is approximately 1.5-3 mm (0.06-0.12 in). Bringing the two into contact also serves to strike an arc, but this can cause contamination of the weld and electrode. Once the arc is struck, the welder moves the torch in a small circle to create a welding pool, the size of which depends on the size of the electrode and the current. While maintaining a constant separation between the electrode and the workpiece, the operator then moves the torch back slightly and tilts it backward about 10-15 degrees from vertical. Filler metal is added manually to the front end of the weld pool as it is needed.

Welders often develop a technique of rapidly alternating between moving the torch forward (to advance the weld pool) and adding filler metal. The filler rod is withdrawn from the weld pool each time the electrode advances, but it is never removed from the gas shield to prevent oxidation of its surface and contamination of the weld. Filler rods composed of metals with low melting temperature, such as aluminium, require that the operator maintain some distance from the arc while staying inside the gas shield. If held too close to the arc, the filler rod can melt before it makes contact with the weld puddle. As the weld nears completion, the arc current is often gradually reduced to prevent the formation of a crater at the end of the weld.

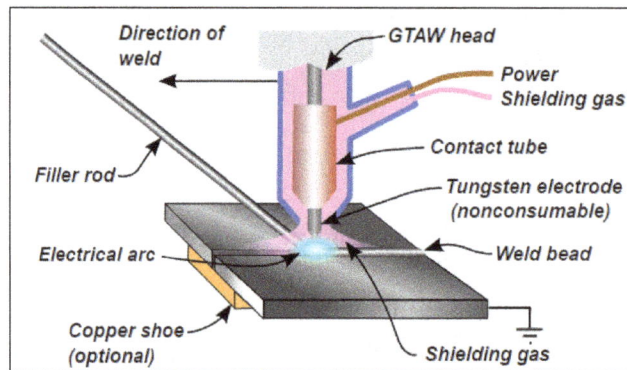

GTAW Weld Area.

Safety

Like other arc welding processes, GTAW can be dangerous if proper precautions are not taken. Welders wear protective clothing, including heavy leather gloves and protective long sleeve jackets, to avoid exposure to extreme heat and flames. Due to the absence of smoke in GTAW, the electric arc can seem brighter than in shielded metal arc welding, making operators especially susceptible to arc eye and skin irritations not unlike sunburn. Helmets with dark face plates are worn to prevent this exposure to ultraviolet light, sometimes featuring a liquid crystal-type face plate that self-darkens upon exposure to high amounts of UV light. Transparent welding curtains, made of a polyvinyl chloride plastic film, are often used to shield nearby workers and bystanders from exposure to the UV light from the electric arc.

Welders are also often exposed to dangerous gases and particulate matter. Shielding gases can displace oxygen and lead to asphyxiation, and while smoke is not produced, the brightness of the arc in GTAW can cause surrounding air to break down and form ozone. Similarly, the brightness and heat can cause poisonous fumes to form from cleaning and degreasing materials. Cleaning operations using these agents should not be performed near the site of welding, and proper ventilation is necessary to protect the welder.

Applications

While the aerospace industry is one of the primary users of gas tungsten arc welding, the process is used in a number of other areas. Many industries use GTAW for welding thin workpieces, especially nonferrous metals. It is used extensively in the manufacture of space vehicles, and is also frequently employed to weld small-diameter, thin-wall tubing. In addition, GTAW is often used to make root or first pass welds for piping of various sizes. In maintenance and repair work, the process is commonly used to repair tools and dies, especially components made of aluminium and magnesium. Because the welds it produces are highly resistant to corrosion and cracking over long time periods, GTAW is the welding procedure of choice for critical welding operations like sealing spent nuclear fuel canisters before burial.

Quality

Among arc welding process, GTAW ranks the highest in terms of the quality of weld produced. Maximum quality is assured by maintaining the cleanliness of the operation—all equipment and materials used must be free from oil, moisture, dirt and other impurities, as these cause weld porosity and consequently a decrease in weld strength and quality. To remove oil and grease, alcohol or similar commercial solvents may be used, while a stainless steel wire brush or chemical process can remove oxides from the surfaces of metals like aluminium. Rust on steels can be removed by first grit blasting the surface and then using a wire brush to remove any embedded grit. These steps are especially important when negative polarity direct current is used, because such a power supply provides no cleaning during the welding process, unlike positive polarity direct current or alternating current. To maintain a clean weld pool during welding, the shielding gas flow should be sufficient and consistent so that the gas covers the weld and blocks impurities in the atmosphere. GTA welding in windy or drafty environments increases the amount of shielding gas necessary to protect the weld, increasing the cost and making the process unpopular outdoors.

Because of GTAW's relative difficulty and the importance of proper technique, skilled operators are employed for important applications. Low heat input, caused by low welding current or high welding speed, can limit penetration and cause the weld bead to lift away from the surface being welded. If there is too much heat input, however, the weld bead grows in width while the likelihood of excessive penetration and spatter increase. Additionally, if the welder holds the welding torch too far from the workpiece, shielding gas is wasted and the appearance of the weld worsens.

If the amount of current used exceeds the capability of the electrode, tungsten inclusions in the weld may result. Known as tungsten spitting, it can be identified with radiography and prevented by changing the type of electrode or increasing the electrode diameter. In addition, if the electrode is not well protected by the gas shield or the operator accidentally allows it to contact the molten metal, it can become dirty or contaminated. This often causes the welding arc to become unstable, requiring that electrode be ground with a diamond abrasive to remove the impurity.

GTAW t-joint weld.

Equipment

The equipment required for the gas tungsten arc welding operation includes a welding torch utilizing a nonconsumable tungsten electrode, a constant-current welding power supply, and a shielding gas source.

GTAW torch with various electrodes, cups,
collets and gas diffusers.

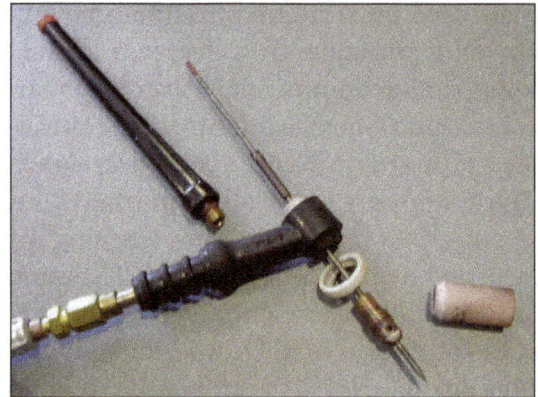

GTAW torch, disassembled.

Welding Torch

GTAW welding torches are designed for either automatic or manual operation and are equipped with cooling systems using air or water. The automatic and manual torches are similar in construction, but the manual torch has a handle while the automatic torch normally comes with a mounting rack. The angle between the centerline of the handle and the centerline of the tungsten electrode, known as the head angle, can be varied on some manual torches according to the preference of the operator. Air cooling systems are most often used for low-current operations (up to about 200 A), while water cooling is required for high-current welding (up to about 600 A). The torches are connected with cables to the power supply and with hoses to the shielding gas source and where used, the water supply.

The internal metal parts of a torch are made of hard alloys of copper or brass in order to transmit current and heat effectively. The tungsten electrode must be held firmly in the centre of the torch with an appropriately sized collet, and ports around the electrode provide a constant flow of shielding gas. The body of the torch is made of heat-resistant, insulating plastics covering the metal components, providing insulation from heat and electricity to protect the welder.

The size of the welding torch nozzle depends on the size of the desired welding arc, and the inside diameter of the nozzle is normally at least three times the diameter of the electrode. The nozzle must be heat resistant and thus is normally made of alumina or a ceramic material, but fused quartz, a glass-like substance, offers greater visibility. Devices can be inserted into the nozzle for special applications, such as gas lenses or valves to control shielding gas flow and switches to control welding current.

Power Supply

Gas tungsten arc welding uses a constant current power source, meaning that the current (and thus the heat) remains relatively constant, even if the arc distance and voltage change. This is important because most applications of GTAW are manual or semiautomatic, requiring that an operator hold the torch. Maintaining a suitably steady arc distance is difficult if a constant voltage power source is used instead, since it can cause dramatic heat variations and make welding more difficult.

The preferred polarity of the GTAW system depends largely on the type of metal being welded. Direct current with a negatively charged electrode (DCEN) is often employed when welding steels, nickel, titanium, and other metals. It can also be used in automatic GTA welding of aluminium or magnesium when helium is used as a shielding gas. The negatively charged electrode generates heat by emitting electrons which travel across the arc, causing thermal ionization of the shielding gas and increasing the temperature of the base material. The ionized shielding gas flows toward the electrode, not the base material, and this can allow oxides to build on the surface of the weld. Direct current with a positively charged electrode (DCEP) is less common, and is used primarily for shallow welds since less heat is generated in the base material. Instead of flowing from the electrode to the base material, as in DCEN, electrons go the other direction, causing the electrode to reach very high temperatures. To help it maintain its shape and prevent softening, a larger electrode is often used. As the electrons flow toward the electrode, ionized shielding gas flows back toward the base material, cleaning the weld by removing oxides and other impurities and thereby improving its quality and appearance.

Alternating current, commonly used when welding aluminium and magnesium manually or semi-automatically, combines the two direct currents by making the electrode and base material alternate between positive and negative charge. This causes the electron flow to switch directions constantly, preventing the tungsten electrode from overheating while maintaining the heat in the base material. This makes the ionized shielding gas constantly switch its direction of flow, causing impurities to be removed during a portion of the cycle. Some power supplies enable operators to use an unbalanced alternating current wave by modifying the exact percentage of time that the current spends in each state of polarity, giving them more control over the amount of heat and cleaning action supplied by the power source. In addition, operators must be wary of rectification, in which the arc fails to reignite as it passes from straight polarity (negative electrode) to reverse polarity (positive electrode). To remedy the problem, a square wave power supply can be used, as can high-frequency voltage to encourage ignition.

GTAW Power Supply.

Electrode

The electrode used in GTAW is made of tungsten or a tungsten alloy, because tungsten has the highest melting temperature among pure metals, at 3,422 °C (6,192 °F). As a result, the electrode is not consumed during welding, though some erosion (called burn-off) can occur. Electrodes can have either a clean finish or a ground finish—clean finish electrodes have been chemically cleaned, while ground finish electrodes have been ground to a uniform size and have a polished surface, making them optimal for heat conduction. The diameter of the electrode can vary between 0.5 millimeter and 6.4 millimeters (0.02–0.25 in), and their length can range from 75 to 610 millimeters (3–24 in).

ISO Class	ISO Colour	AWS Class	AWS Colour	Alloy
WP	Green	EWP	Green	None
WC20	Gray	EWCe-2	Orange	~2% CeO_2
WL10	Black	EWLa-1	Black	~1% La_2O_3
WL15	Gold	EWLa-1.5	Gold	~1.5% La_2O_3
WL20	Sky-blue	EWLa-2	Blue	~2% La_2O_3
WT10	Yellow	EWTh-1	Yellow	~1% ThO_2
WT20	Red	EWTh-2	Red	~2% ThO_2
WT30	Violet			~3% ThO_2
WT40	Orange			~4% ThO_2
WY20	Blue			~2% Y_2O_3
WZ3	Brown	EWZr-1	Brown	~0.3% ZrO_2
WZ8	White			~0.8% ZrO_2

A number of tungsten alloys have been standardized by the International Organization for Standardization and the American Welding Society in ISO 6848 and AWS A5.12, respectively, for use in GTAW electrodes, and are summarized in the adjacent table. Pure tungsten electrodes (classified as WP or EWP) are general purpose and low cost electrodes. Cerium oxide (or ceria) as an alloying element improves arc stability and ease of starting while decreasing burn-off. Using an alloy of lanthanum oxide (or lanthana) has a similar effect. Thorium oxide (or thoria) alloy electrodes were designed for DC applications and can withstand somewhat higher temperatures while providing many of the benefits of other alloys. However, it is somewhat radioactive, and as a replacement, electrodes with larger concentrations of lanthanum oxide can be used. Electrodes containing zirconium oxide (or zirconia) increase the current capacity while improving arc stability and starting and increasing electrode life. In addition, electrode manufacturers may create alternative tungsten alloys with specified metal additions, and these are designated with the classification EWG under the AWS system.

Filler metals are also used in nearly all applications of GTAW, the major exception being the welding of thin materials. Filler metals are available with different diameters and are made of a variety of materials. In most cases, the filler metal in the form of a rod is added to the weld pool manually, but some applications call for an automatically fed filler metal, which often is stored on spools or coils.

Shielding Gas

As with other welding processes such as gas metal arc welding, shielding gases are necessary in GTAW to protect the welding area from atmospheric gases such as nitrogen and oxygen, which can cause fusion defects, porosity, and weld metal embrittlement if they come in contact with the electrode, the arc, or the welding metal. The gas also transfers heat from the tungsten electrode to the metal, and it helps start and maintain a stable arc.

The selection of a shielding gas depends on several factors, including the type of material being welded, joint design, and desired final weld appearance. Argon is the most commonly used shielding gas for GTAW, since it helps prevent defects due to a varying arc length. When used with alternating current, the use of argon results in high weld quality and good appearance. Another common shielding gas, helium, is most often used to increase the weld penetration in a joint, to increase the welding speed, and to weld metals with high heat conductivity, such as copper and aluminium. A significant disadvantage is the difficulty of striking an arc with helium gas, and the decreased weld quality associated with a varying arc length.

Argon-helium mixtures are also frequently utilized in GTAW, since they can increase control of the heat input while maintaining the benefits of using argon. Normally, the mixtures are made with primarily helium (often about 75% or higher) and a balance of argon. These mixtures increase the speed and quality of the AC welding of aluminium, and also make it easier to strike an arc. Another shielding gas mixture, argon-hydrogen, is used in the mechanized welding of light gauge stainless steel, but because hydrogen can cause porosity, its uses are limited. Similarly, nitrogen can sometimes be added to argon to help stabilize the austenite in austentitic stainless steels and increase penetration when welding copper. Due to porosity problems in ferritic steels and limited benefits, however, it is not a popular shielding gas additive.

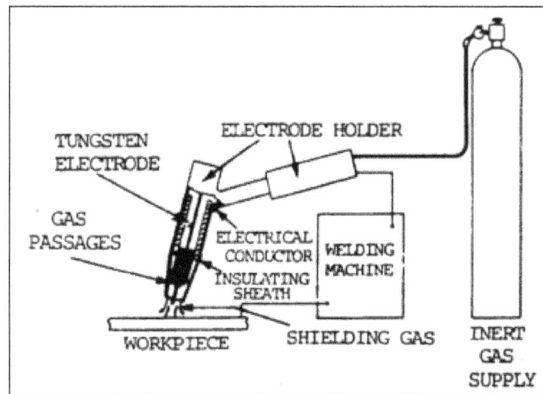

GTAW system setup.

Materials

Gas tungsten arc welding is the most commonly used to weld stainless steel and nonferrous materials, such as aluminium and magnesium, but it can be applied to nearly all metals, with notable exceptions being lead and zinc. Its applications involving carbon steels are limited not because of process restrictions, but because of the existence of more economical steel welding techniques, such as gas metal arc welding and shielded metal arc welding. Furthermore, GTAW can be performed in a variety of other-than-flat positions, depending on the skill of the welder and the materials being welded.

Aluminium and Magnesium

Aluminium and magnesium are most often welded using alternating current, but the use of direct current is also possible, depending on the properties desired. Before welding, the work area should be cleaned and may be preheated to 175 to 200 °C (350 to 400 °F) for aluminium or to a maximum of 150 °C (300 °F) for thick magnesium workpieces to improve penetration and increase travel speed. AC current can provide a self-cleaning effect, removing the thin, refractory aluminium oxide (sapphire) layer that forms on aluminium metal within minutes of exposure to air. This oxide layer must be removed for welding to occur. When alternating current is used, pure tungsten electrodes or zirconiated tungsten electrodes are preferred over thoriated electrodes, as the latter are more likely to "spit" electrode particles across the welding arc into the weld. Blunt electrode tips are preferred, and pure argon shielding gas should be employed for thin workpieces. Introducing helium allows for greater penetration in thicker workpieces, but can make arc starting difficult.

Direct current of either polarity, positive or negative, can be used to weld aluminium and magnesium as well. Direct current with a negatively charged electrode (DCEN) allows for high penetration, and is most commonly used on joints with butting surfaces, such as square groove joints. Short arc length (generally less than 2 mm or 0.07 in) gives the best results, making the process better suited for automatic operation than manual operation. Shielding gases with high helium contents are most commonly used with DCEN, and thoriated electrodes are suitable. Direct current with a positively charged electrode (DCEP) is used primarily for shallow welds, especially those with a joint thickness of less than 1.6 millimeters (0.06 in). While still important, cleaning is less essential for DCEP than DCEN, since the electron flow from the workpiece to the electrode helps maintain a clean weld. A large, thoriated tungsten electrode is commonly used, along with a pure argon shielding gas.

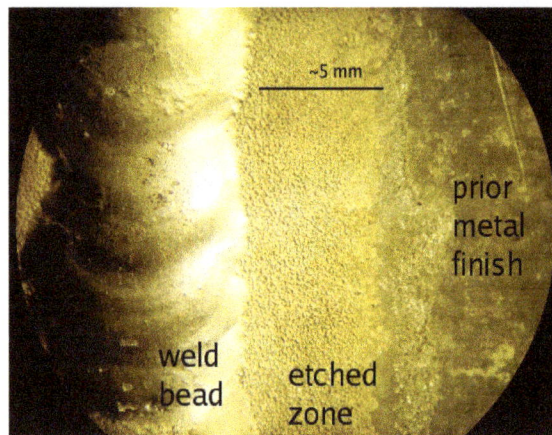

A TIG weld showing an accentuated AC etched zone.

Steels

For GTA welding of carbon and stainless steels, the selection of a filler material is important to prevent excessive porosity. Oxides on the filler material and workpieces must be removed before welding to prevent contamination, and immediately prior to welding, alcohol or acetone should be used to clean the surface. Preheating is generally not necessary for mild steels less than one inch thick, but low alloy steels may require preheating to slow the cooling process and prevent the formation of martensite in the heat-affected zone. Tool steels should also be preheated to prevent cracking in the heat-affected zone. Austenitic stainless steels do not require preheating, but martensitic and ferritic chromium stainless steels do. A DCEN power source is normally used, and thoriated electrodes, tapered to a sharp point, are recommended. Pure argon is used for thin workpieces, but helium can be introduced as thickness increases.

Dissimilar Metals

Welding dissimilar metals often introduces new difficulties to GTA welding, because most materials do not easily fuse to form a strong bond. However, welds of dissimilar materials have numerous applications in manufacturing, repair work, and the prevention of corrosion and oxidation. In some joints, a compatible filler metal is chosen to help form the bond, and this filler metal can be the same as one of the base materials (for example, using a stainless steel filler metal with stainless steel and carbon steel as base materials), or a different metal (such as the use of a nickel filler metal for joining steel and cast iron). Very different materials may be coated or "buttered" with a material compatible with a particular filler metal, and then welded. In addition, GTAW can be used in cladding or overlaying dissimilar materials.

When welding dissimilar metals, the joint must have an accurate fit, with proper gap dimensions and bevel angles. Care should be taken to avoid melting excessive base material. Pulsed current is particularly useful for these applications, as it helps limit the heat input. The filler metal should be added quickly, and a large weld pool should be avoided to prevent dilution of the base materials.

Process Variations

Pulsed-Current

In the pulsed-current mode, the welding current rapidly alternates between two levels. The higher current state is known as the pulse current, while the lower current level is called the background current. During the period of pulse current, the weld area is heated and fusion occurs. Upon dropping to the background current, the weld area is allowed to cool and solidify. Pulsed-current GTAW has a number of advantages, including lower heat input and consequently a reduction in distortion and warpage in thin workpieces. In addition, it allows for greater control of the weld pool, and can increase weld penetration, welding speed, and quality. A similar method, manual programmed GTAW, allows the operator to program a specific rate and magnitude of current variations, making it useful for specialized applications.

Dabber

The dabber variation is used to precisely place weld metal on thin edges. The automatic process replicates the motions of manual welding by feeding a cold filler wire into the weld area and dabbing (or oscillating) it into the welding arc. It can be used in conjunction with pulsed current, and is used to weld a variety of alloys, including titanium, nickel, and tool steels. Common applications include rebuilding seals in jet engines and building up saw blades, milling cutters, drill bits, and mower blades.

Gas Metal Arc Welding

Gas metal arc welding (GMAW), sometimes referred to by its subtypes as metal inert gas (MIG) welding or metal active gas (MAG) welding, is an electric arc welding process where the heat for welding is produced by an arc between a continuously fed, consumable filler metal electrode and the work. The shielding of the molten weld pool and the arc is obtained from an externally supplied gas or gas mixture.

The gas metal arc welding process uses the heat of an electric arc produced between a bare electrode and the part to be welded. The electric arc is produced by electric current passing through an ionized gas. The gas atoms and molecules are broken up and ionized by losing electrons and leaving a positive charge. The positive gas ions then flow from the positive pole to the negative pole, and the electrons flow from the negative pole to the positive pole. About 95% of the heat is carried by the electrons, and the rest is carried by the positive ions. The heat of the arc melts the surface of the base metal and the electrode. The molten weld metal, heated weld zone, and the electrode are shielded from the atmosphere by a shielding gas supplied through the welding gun. The molten electrode filler metal transfers across the arc and into the weld puddle. This process produces an arc with more intense heat than most of the arc welding processes.

The arc is struck by starting the wire feed, which causes the electrode wire to touch the workpiece and initiate the arc. Normally, arc travel along the work is not started until a weld puddle is formed. The gun then moves along the weld joint manually or mechanically so that the adjoining edges are joined. The weld metal solidifies behind the arc in the joint and completes the welding process.

Arc Systems

The gas metal arc welding process may be operated on both constant voltage and constant current power sources. Any welding power source can be classified by its voltampere characteristics as either a constant voltage (also called constant potential) or constant current (also called variable voltage) type although there are some machines that can produce both characteristics. Constant voltage power sources are preferred for a majority of gas metal arc welding applications.

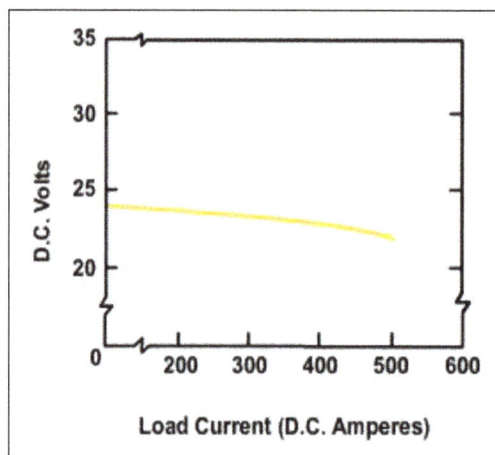

Volt-amp curve.

In the constant voltage arc system, the voltage delivered to the arc is maintained at a relatively constant level, which gives a flat or nearly flat volt-ampere curve. This type of power source is widely used for the processes that require a continuously fed bare wire electrode. In this system, the arc length is controlled by setting the voltage level on the power source, and the welding current is controlled by setting the wire feed speed.

Most machines have a fixed slope that is built in for a certain type of gas metal arc welding. Some constant voltage welding machines are equipped with a slope control that is used to change the slope of the volt-ampere curve. Figure shows different slopes obtained from one power source. The slope has the effect of limiting the amount of short-circuiting current that the power supply can deliver. This is the current available from the power source on the short circuit between the electrode wire and the work.

A slope control is not required but is best when welding with small diameter wire and low current levels. The short-circuit current determines the amount of pinch force available on the electrode. The pinch forces cause the molten electrode tip to neck down so that the droplet will separate from the solid electrode. The flatter the slope of the volt-ampere curve, the higher the shortcircuit current and the pinch force. The steeper the slope the lower the short circuit current and pinch force. The pinch force is important because it affects the way the droplet detaches from the tip of the electrode wire, which also affects the arc stability in short-circuiting transfer. When a high shortcircuit and pinch force are caused by a flat slope, excessive spatter is created. When a very low short circuit current and pinch force are caused by a steep slope, the electrode wire tends to freeze in the weld puddle or pile upon the work piece. When the proper amount of short-circuit current is used, very little spatter with a smooth electrode tip is created.

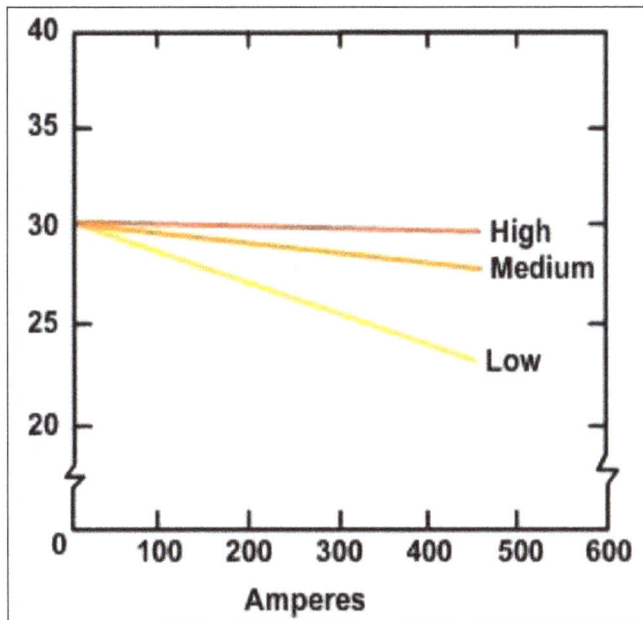

Volt-amp slopes.

The inductance of the power supply also has an effect on the arc stability. When loads on the power supply change, the output current will fluctuate, taking time to find its new level. The rate of current change is determined by the inductance of the power supply. The rate of the welding current buildup and pinch force buildup increases with the current, which is also affected by the inductance in the circuit. Increasing the inductance will reduce the rate of current rise and the pinch force. (In short-circuiting welding, increasing the inductance will increase the arc time between short-circuit and decrease the frequency of short-circuiting, thereby reducing the amount of spatter). Increased arc time or inductance produces a flatter and smoother weld bead as well as a more fluid weld puddle. Too much inductance will cause more difficult arc starting.

The constant current (CC) arc system provides a nearly constant welding current to the arc, which gives a drooping volt-ampere characteristic. This arc system is used with the shielded metal arc welding and gas tungsten arc welding processes. The welding current is set by a dial on the machine, and the welding voltage is controlled by the arc length held by the welder. This system is necessary for manual welding because the welder cannot hold a constant arc length, which causes only small variations in the welding current. When gas metal arc welding is done with a constant current system, a special voltage sensing wire feeder is used to maintain a constant arc length.

For any power source, the voltage drop across the welding arc is directly dependent on the arc length. An increase in the arc length results in a corresponding increase in the arc voltage, and a decrease in the arc length results in a corresponding decrease in the arc voltage. Another important relationship exists between the welding current and the melt off rate of the electrode. With low current, the electrode melts off slower and the metal is deposited slower. This relationship between welding current and wire feed speed is definite, based on the wire size, shielding gas, and type of filler metal; a faster wire feed speed will give a higher welding current.

In the constant voltage system, instead of regulating the wire to maintain a constant arc length, the wire is fed into the arc at a fixed speed, and the power source is designed to melt off the wire at the same speed. The self-regulating characteristic of a constant voltage power source comes about by the ability of this type of power source to adjust its welding current to maintain a fixed voltage across the arc.

With the constant current arc system with a voltage sensing wire feeder, the welder would change the wire feed speed as the gun is moved toward or away from the weld puddle. Since the welding current remains the same, the burn-off rate of the wire is unable to compensate for the variations in the wire feed speed, which allows stubbing or burning back of the wire into the contact tip to occur. To lessen this problem, a special voltage sensing wire feeder is used which regulates the wire feed speed to maintain a constant voltage across the arc.

The constant voltage system is preferred for most applications, particularly for small diameter wire. With smaller diameter electrodes, the voltage sensing system is often not able to react fast enough to feed at the required burn--off rate, resulting in a higher instance of burnback into the contact tip of the gun.

The figure shows a comparison of the volt-ampere curves for the two arc systems. This shows that for these particular curves, when a normal arc length is used, the current and voltage level is the same for both the constant current and constant voltage systems. For a long arc length, there is a slight drop in the welding current for the constant current machine and a large drop in the current for a constant voltage machine. For constant voltage power sources, the volt-ampere curve shows that when the arc length shortens slightly, a large increase in welding current occurs. This results in an increased burn-off rate which brings the arc length back to the desired level. Under this system, changes in the wire feed speed caused by the welder are compensated for electrically by the power source. The constant current system is sometimes used, especially for welding aluminium and magnesium because the welder can vary the current slightly by changing the arc length. This varies the depth of penetration and the amount of heat input. With aluminium and magnesium, preheating the wire is not desirable.

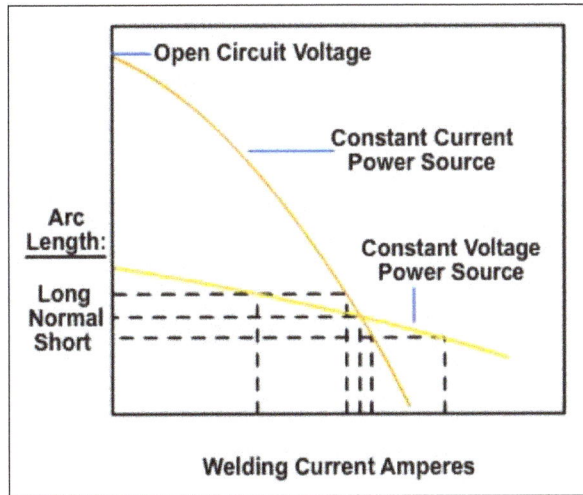

Volt-amp curves.

Metal Transfer

The types of arcs obtainable and the different modes of gas metal arc welding are determined by the type of metal transfer. The four modes of welding are the short circuiting, globular, spray, and pulsed arc metal transfer. Each mode has its own advantages and applications. The type of metal transfer is determined by the welding current, shielding gas, and welding voltage.

Short Circuiting Transfer

At the beginning of the short-circuiting arc cycle, the end of the electrode wire melts into a small globule which moves toward the weld puddle. When the tip of this globule comes in contact with the workpiece, the arc is momentarily extinguished. When the wire touches the workpiece, the current increases because a short circuit is created. The current increases to the point that the molten globule is pinched off and the arc is re-ignited. This cycle then repeats itself, occurring approximately 20 to 200 times a second depending on the current level and the power supply. The filler metal is transferred to the weld puddle only during the period when the electrode is in contact with the work. No filler metal is transferred across the arc.

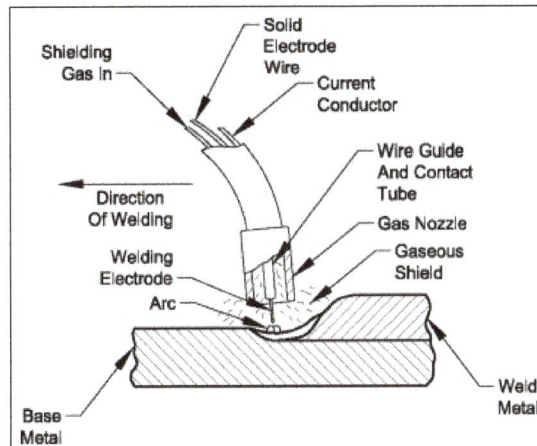

Short-circuiting transfer.

Short-circuiting transfer applies the lowest welding currents and voltages used with gas metal arc welding, which produces low heat input. The type of shielding gas used has very little effect on this type of transfer but most gas metal arc welding done in this mode employs a CO_2 shielding gas. This type of metal transfer produces a small, fastfreezing weld pool, usually with some small, fine spatter. Because of this, this mode is well suited for joining thin sections of metal by welding in the vertical, horizontal, and overhead positions, and for filling large root openings.

Globular Transfer

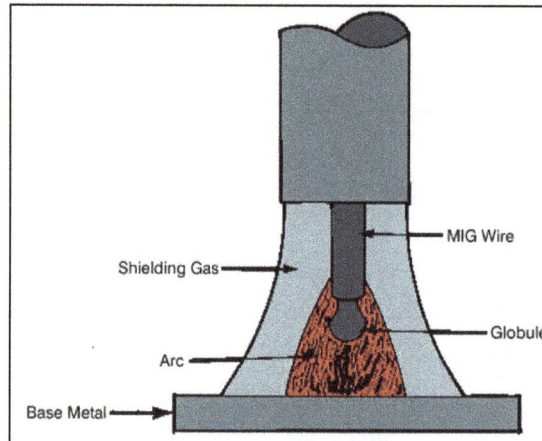

Globular transfer.

The globular transfer cycle starts when a droplet forms on the end of the electrode wire. The molten droplet grows in size until it is larger than the diameter of the electrode. The droplet then detaches from the end of the electrode and transfers across the arc due to the force of gravity. Globular transfer is shown in figure.

Globular transfer occurs at relatively low operating currents and voltages but higher than those used to obtain short-circuiting transfer. It can occur with all types of shielding gases, but with gases other than CO_2 it generally occurs at current and voltage levels toward the bottom of the operating range. With CO_2 shielding gas, globular transfer will take place at most operating current and voltage levels. Because of the large droplet size and the dependence on gravity to transfer the filler metal, this mode of gas metal arc welding is not suitable for many out-of-position welding applications, especially overhead welding where the droplets tend to fall into the nozzle of the welding gun. Globular transfer is also characterized by a less stable arc and higher amounts of spatter. The arc is less stable because it will shift around and move to the part of the droplet that is closest to the weld puddle, (electric current will always try to take the shortest path). The arc will wave around on the end of the droplet, creating more spatter.

Spray Transfer

The spray transfer cycle begins when the end of the electrode tapers down to a point. Small droplets are formed and electromagnetically pinched off at the tapered point of the electrode tip. The droplets are smaller than the diameter of the electrode and detach much more rapidly than in globular transfer. The rate of transfer can vary from less than one hundred times a second up to several hundred times a second. The arc is also more directional than in the globular mode. Spray transfer is shown in figure.

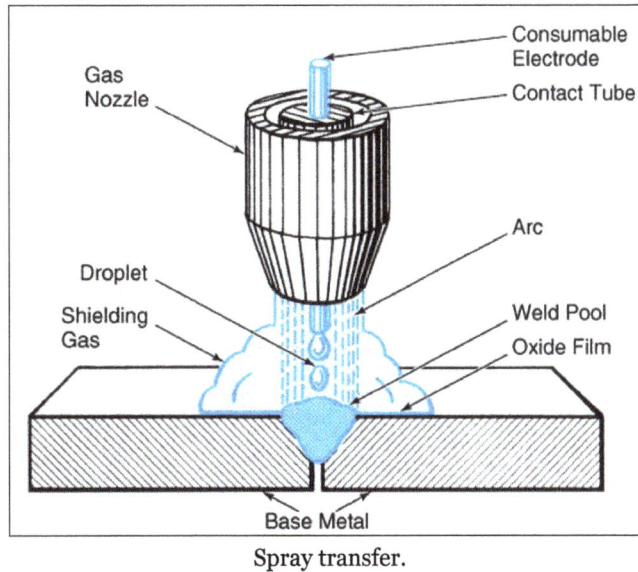

Spray transfer.

Spray transfer is generally associated with the higher amperage and voltage levels and occurs with argon or argon-rich shielding gases. The spray transfer mode is best adapted for welding thick sections because of the higher welding currents. Spray transfer produces a very stable arc that is well adapted for out-of-position as well as flat position welding. When welding out-ofposition, operators need to consider how the high voltage and current levels used may produce a weld puddle that is difficult to control. This mode also produces the least amount of spatter.

Pulsed Current Transfer

To overcome the work thickness and welding position limitations of spray transfer, specially designed power supplies have been developed. These machines produce controlled wave forms and frequencies that "pulse" the welding current at regularly spaced intervals. They provide two levels of current: one a constant, low background current which sustains the arc without providing enough energy to cause drops to form on the wire tip; the other is a superimposed pulsing current with amplitude greater than the transition current necessary for spray transfer. During this pulse, one or more drops are formed and transferred. The frequency and amplitude of the pulses control the energy level of the arc, and therefore the rate at which the wire melts. By reducing the average arc energy and the wire-melting rate, pulsing makes the desirable features of spray transfer available for joining sheet metals and welding thick metals in all positions.

Equipment for Welding

The basic design of a GMAW system is shown in and includes four principal components:

- Power source.

- Wire drive and accessories (drive rolls, guide tubes, reel stand, etc.).

- GMAW gun and cable assembly designed to deliver the shielding gas and the electrode to the arc.

- Shielding gas apparatus and accessories.

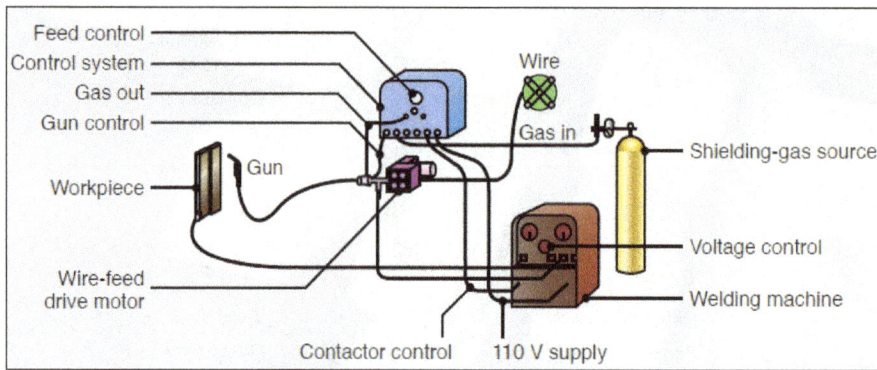

Equipment for gas metal arc welding.

Power Sources

The purpose of the power source or welding machine is to provide the electric power of the proper current and voltage to maintain a welding arc. Many power sources operate on 200, 230, 460, or 575 volt input electric power. The power sources operate on singlephase or three-phase input power with a frequency of 50 or 60 Hz.

Power Source Duty Cycle

The duty cycle of a power source is defined as the ratio of arc time to total time. Most power sources used for gas metal arc welding have a duty cycle of 100%, which indicates that they can be used to weld continuously. Some machines used for this process have duty cycles of 60%, which means that they can be used to weld six of every ten minutes. In general, these lower duty cycle machines are the constant current type that are used in plants where the same machines are also used for shielded metal arc welding and gas tungsten arc welding. Some of the smaller constant voltage welding machines have a 60% duty cycle.

Types of Current

Most gas metal arc welding is done using steady direct current. Steady direct current can be connected in one of two ways: electrode positive (reverse polarity DCEP) and electrode negative (straight polarity DCEN). The electrically charged particles flow between the tip of the electrode and the work. The electrode positive connection is used for almost all welding applications of this process. It gives better penetration than electrode negative and can be used to weld all metals. Electrode negative is sometimes used when a minimum amount of penetration is desired.

Pulsed direct current is used for applications where good penetration and reduced heat input are required. Pulsed current occurs when the welding current is operated at one level for a set period of time, switches to another level for a time, and then repeats the cycle. The pulsing action can be provided from one power source or combining the outputs of two power sources working at two current levels. The welding current varies from as low as 20 amps at 18 volts up to as high as 750 amps at 50 volts, and the frequency of pulsing can be varied. When using pulsed current, welding thinner sections is more practical than when using steady direct current in the spray transfer mode, because there is less heat input, which reduces the amount of distortion.

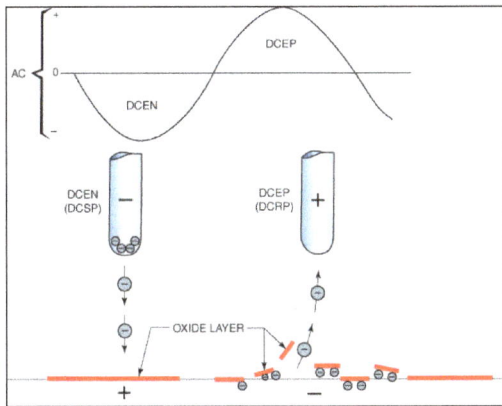

Particle Flow for Dcep And Dcen.

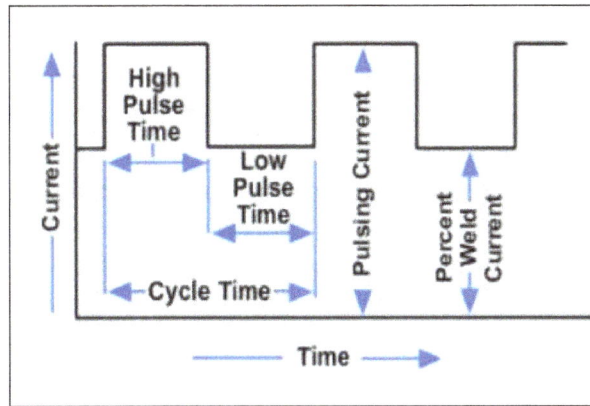

Pulsed current terminology.

Types of Power Sources

Many types of direct current power sources may be used for gas metal arc welding, including engine-driven generators (rotating) and transformer-rectifiers (static). Inverters are included in the static category.

Generator Welding Machines

A generator welding machine can be powered by an electric motor for shop use or by an internal combustion engine (gas or diesel) for field use. Engine-driven welders can have either water- or air-cooled engines, and many of them provide auxiliary power as well Many of the engine-driven generators used for gas metal arc welding in the field are combination constant current-constant voltage types. These are popular for applications such as pipe welding so that both shielded metal arc welding and gas metal arc welding can be done using the same power source. The motor-driven generator welding machines are becoming less popular and are being replaced by transformer-rectifier welding machines. Motor-driven generators produce a very stable arc, but they are noisier and more expensive, consume more power, and require more maintenance than transformer-rectifier machines.

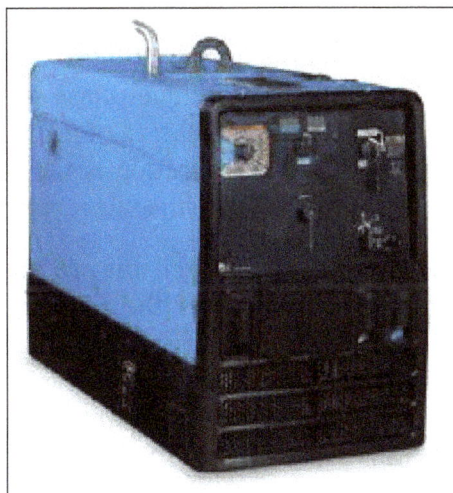

Engine-driven power source.

Transformer-Rectifier Welding Machines

The more popular welding machines used for gas metal arc welding are the transformer-rectifiers. A method of supplying direct current to the arc other than the use of a rotating generator is by adding a rectifier to a basic transformer circuit. A rectifier is an electrical device which changes alternating current into direct current. These machines are more efficient electrically than motorgenerator welding machines, they respond faster when arc conditions change, and they provide quieter operation. There are two basic types of transformer-rectifier welding machines: those that operate on singlephase input power and those that operate on three-phase input power The single-phase transformer-rectifier machines provide DC current to the arc and a constant current volt-ampere characteristic. These machines are not as popular as three-phase transformer-rectifier welding machines for gas metal arc welding. When using a constant current power source, a special variable speed or voltage sensing wire feeder must be used to keep the current level uniform.

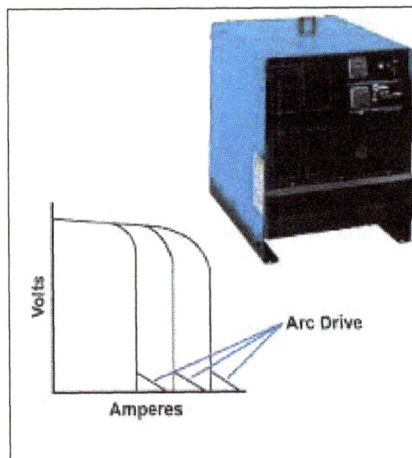

Three-phase constant voltage.

The single-phase transformer-rectifier machines provide DC current to the arc and a constant current volt-ampere characteristic. These machines are not as popular as three-phase transformer-rectifier welding machines for gas metal arc welding. When using a constant current power source, a special variable speed or voltage sensing wire feeder must be used to keep the current level uniform.

Machines used for shielded metal arc welding and gas tungsten arc welding can be adapted for use with gas metal arc welding. A limitation of the single-phase system is that the power required by the single-phase input power may create an unbalance of the power supply lines, which is objectionable to most power companies. Another limitation is that short-circuiting metal transfer cannot be used with this type of power source. These machines normally have a duty cycle of 60%.

One of the most widely used types of power sources for this process is the three-phase transformer rectifier. These machines produce DC current for the arc and most have a constant voltage volt-ampere characteristic. When using these machines, a constant speed wire feeder is normally employed. This type of wire feeder maintains a constant wire feed speed with slight changes in welding current. The three-phase input power gives these machines a more stable arc than single-phase input power, and avoids the line unbalance that occurs with the single-phase machines. Many of these machines also use solid-state controls for the welding. A solid-state machine will produce the flattest volt-ampere curve of the different constant voltage power sources.

Inverter Power Sources

The inverter machine is different from a transformer-rectifier. The inverter will rectify 60 Hz alternating line current, utilize a chopper circuit to produce a high frequency alternating current, reduce that voltage with an AC transformer, and finally rectify that to obtain the required direct current output. Changing that alternating current frequency to a much higher frequency allows a greatly reduced size of transformer and reduced transformer losses as well Inverter circuits control the output power using the principle of time ratio control (TRC). The solid-state devices (semiconductors) in an inverter act as switches; they are either switched "on" and conducting, or they are switched "off" and blocking. This operation of switching "on" and "off" is sometimes referred to as switch mode operation. TRC is the regulation of the "on" and "off" time of the switches to control the output. Faster response times are generally associated with the higher switching and control frequencies, resulting in more stable arcs and superior arc performance. However, other variables, such as length of weld cables, must be considered since they may affect the power supply performances.

Inverter power source.

Controls

The controls for this process are located on the front of the welding machine, on the welding gun, and on the wire feeder or a control box.

The welding machine controls for a constant voltage machine are an on-off switch, a voltage control, and sometimes a switch to select the polarity of direct current. The voltage control can be a single knob, or it can have a top switch for setting the voltage range and a fine voltage control knob. Other controls are sometimes present such as a switch for selecting CC (constant current) or CV (constant voltage) output on combination machines or a switch for a remote control. On the constant current welding machines there is an on-off switch, a current level control knob, and sometimes a knob or switch for selecting the polarity of direct current.

The trigger or switch on the welding gun is a remote control that is used by the welder in semiautomatic welding to stop and start the welding current, wire feed, and shielding gas flow.

For semiautomatic welding, a wire feed speed control is normally part of the wire feeder assembly or close by. The wire feed speed sets the welding current level on a constant voltage machine. For machine or automatic welding, a separate control box is often used to control the wire feed speed. On the wire feeder control box, there may also be switches to turn the control on and off and gradually feed the wire up and down.

Other controls for this process are used for special applications, especially when using a programmable power source. A couple of examples are items such as timers for spot welding and pulsation.

Wire Feeders

The electrode feed unit (wire feeder) provides the power for driving the electrode through the cable and gun and to the work. There are several different electrode feed units available, but the best type of system depends on the application. Most of the electrode feed units used for gas metal arc welding are the constant speed type which are used with constant voltage power sources. This means that the wire feed speed is set before welding. The wire feed speed controls the amount of welding current.

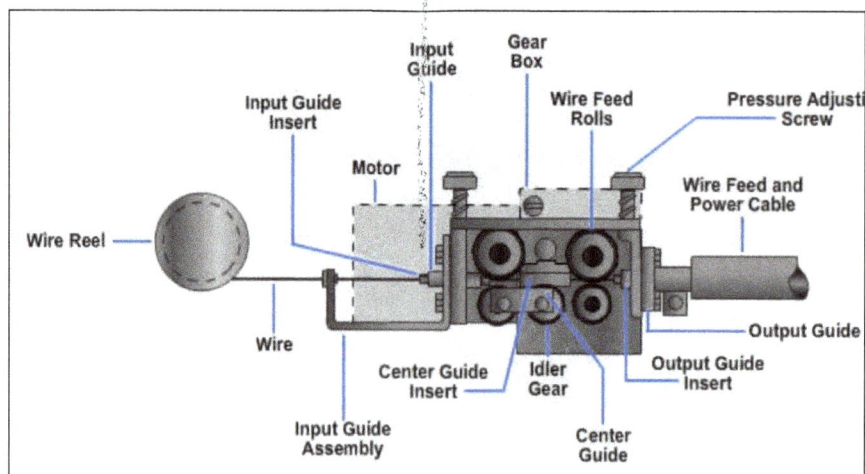

Wire feed assembly.

Variable speed or voltage sensing wire feeders are used with constant current power sources. With a variable speed wire feeder, a voltage sensing circuit is used to maintain the desired arc length by varying the wire feed speed. Variations in the arc length increase or decrease the wire feed speed. The wire-feed speed is measured in inches per minute (ipm). For a specific amperage setting, a high wire-feed speed results in a short arc, whereas a low speed produces a long arc. Therefore, you would use higher speeds for overhead welding than for flat-position welding.

An electrode feed unit consists of an electric motor connected to a gearbox with drive rolls in it. Systems may have two or four feed rolls in the gearbox. In a four roll system, the lower two rolls drive the wire and have a circumferential "V" groove in them, depending on the type and size of wire being fed. Figure shows several of the most common drive rolls and their uses.

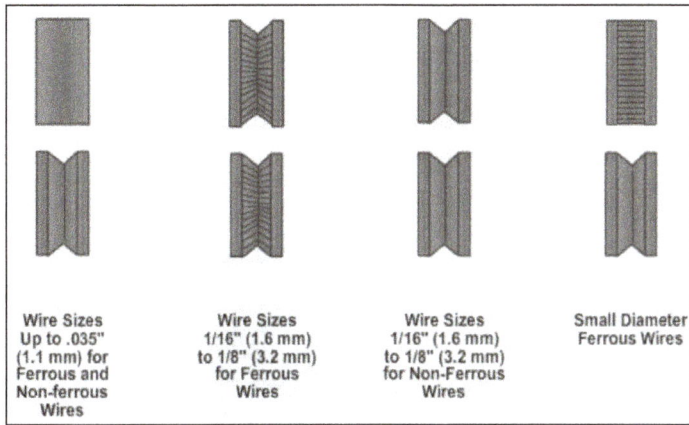

| Wire Sizes Up to .035" (1.1 mm) for Ferrous and Non-ferrous Wires | Wire Sizes 1/16" (1.6 mm) to 1/8" (3.2 mm) for Ferrous Wires | Wire Sizes 1/16" (1.6 mm) to 1/8" (3.2 mm) for Non-Ferrous Wires | Small Diameter Ferrous Wires |

Common types of drive rolls and their uses.

Wire feed systems may be of the push, pull, or push-pull types depending on the type and size of the electrode wire and the distance between the welding gun and the coil or spool of electrode wire. The push type is the wire feeding system most commonly used for steels. It consists of the wire being pulled from the wire feeder by the drive rolls and then being pushed into the flexible conduit and through the gun. The length of the conduit can be up to about 12 ft. (3.7m) for steel wire and 6 ft. (1.8m) for aluminium wire.

A typical push wire feeder is shown in figure. This solid-state wire feeder has the wire feeder control box and the wire reel support mounted with the wire feed motor and gear box.

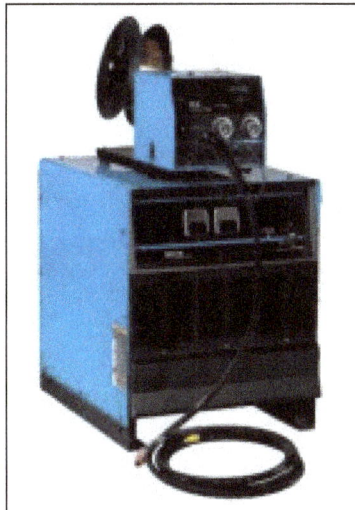

Solid-state control wire feeder and wire support.

Pull type wire feeders have the drive rolls attached to the welding gun. This type of system works best for feeding wires up to about .045 in. (1.1mm) in diameter with a hand-held welding gun. Most machine and automatic welding stations also use this type of system.

The push-pull system is particularly well suited for use with low strength wires such as aluminium and when driving wires long distances. This system can use synchronous drive motors to feed the electrode wire, which makes it good for soft wires and long distances. The wire feeding system shown in uses the standard feeder as the drive motor (push) and the gun as a slave motor (pull).

Standard push-pull wire feeding system.

Welding Guns

A typical GMAW gun is shown in the figure. The welding gun transmits the welding current to the electrode. Because the wire is fed continuously, a sliding electrical contact is used. The welding current is passed to the electrode through a copper base alloy contact tube. The contact tubes have various hole sizes, depending on the diameter of the electrode wire. The gun also has a gas supply connection and a nozzle to direct the shielding gas around the arc and weld puddle. To prevent overheating of the welding gun, cooling is required to remove the heat generated. Shielding gas or water circulating in the gun, or both are used for cooling. An electrical switch is used to start and stop the electrode feeding, welding current, and shielding gas flow. This is located on the gun in semiautomatic welding and separately on machine welding heads.

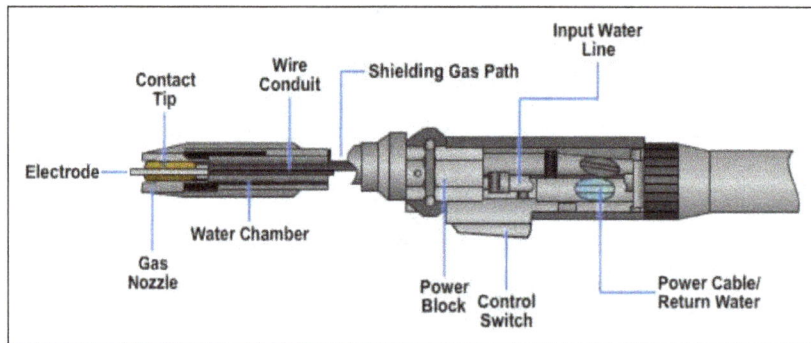

Cross-sectional view of a welding gun.

Semiautomatic Guns

The hand-held semiautomatic guns usually have a curved neck, which makes them flexible, and a curved handle that adds comfort and balance. The gun is attached to the service lines which include the power cable, water hose, gas hose, and wire conduit or liner. The guns have metal nozzles, which have orifice diameters from 3/8 to 7/8 in. (10- 22 mm), depending on the welding requirements, to direct the shielding gas to the arc and weld puddle.

Welding guns are either air-cooled or water-cooled. The choice between the guns is based on the type of shielding gas, amount of welding current, voltage, joint design, and the shop practice. A water-cooled gun is similar to an air-cooled gun except that ducts have been added that permit the cooling water to circulate around the contact tube and nozzle. Water-cooled guns provide more efficient cooling of the gun.

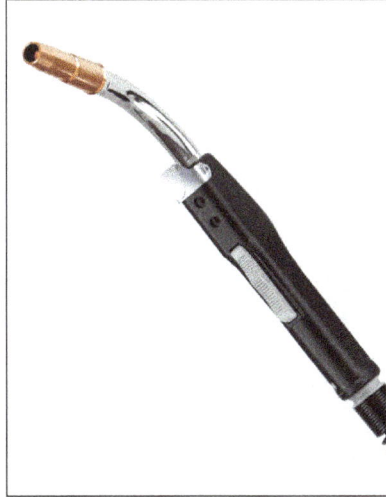

Semi-automatic.

Air-cooled guns are employed for applications where water is not readily available. These are actually cooled by the shielding gas. The guns are available for service up to 600 amperes used intermittently with a CO_2 shielding gas. These guns are usually limited to 50% of the CO_2 rating with argon or helium. CO_2 cools the welding gun, where argon or helium do not. Water-cooling permits the gun to operate continuously at the rated capacity with lower heat buildup. Water-cooled guns are generally used for applications requiring between 200 and 750 amperes. Air-cooled guns of the same capacity as water-cooled guns are heavier but they are easier to manipulate in confined spaces or for out-of position applications because there are fewer cables.

There are three general types of guns available. The one shown in figure has the electrode wire fed through a flexible conduit from a remote wire feeder. The conduit is generally 10 to 15 feet due to the wire feeding limitations of a push type wire feeding system. Figure shows the second type of welding gun, which has a self contained wire feeding mechanism and electrode wire supply. This wire supply is in the form of a 1 lb. (.45 kg) spool.

The third type of gun has a wire feed motor on the gun, and the wire is fed through a conduit from a remote wire feed supply. This system has a pull type wire feeder and can use longer length conduits.

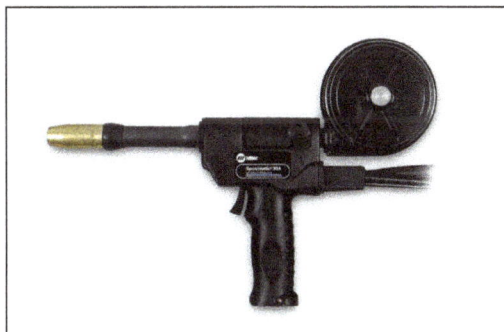

Spool gun.

Machine Welding Guns

The machine welding guns use the same basic design principles and features as the semiautomatic welding guns. These guns have capacities up to 1200 amperes and are generally water-cooled because of the higher amperages and duty cycles required. The gun is mounted directly below the wire feeder. Large diameter wires up to 1/4 in. (6.4 mm) are often used. Figure shows a GMAW control panel for a machine welding gun system.

Shielding Gas Equipment

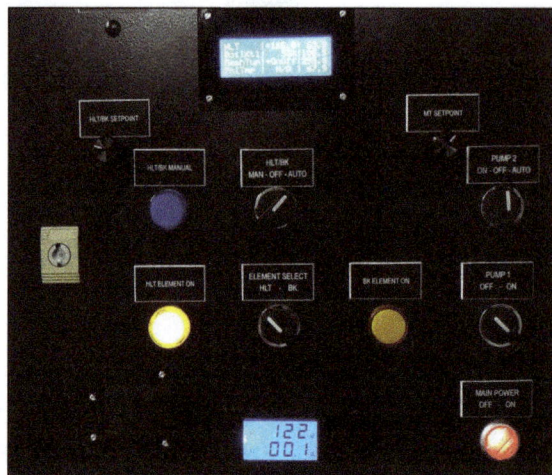

Control panel.

The shielding gas system used in gas metal arc welding consists of a gas supply source, a gas regulator, a flowmeter, control valves, and supply hoses to the welding gun.

The shielding gases are supplied in liquid form when they are in storage tanks with vaporizers or in a gas form in high-pressure cylinders. An exception to this is carbon dioxide. When put in high-pressure cylinders, it exists in both the liquid and gas forms. The bulk storage tank system is used when there are large numbers of welding stations using the same type of shielding gas in large quantities. For applications where there are large numbers of welding stations but relatively low gas usage, a manifold system is often used. This consists of several highpressure cylinders connected to a manifold which then feeds a single line to the welding stations. Individual high-pressure cylinders are used when the amount of gas usage is low, when there are few welding stations, or when portability is required.

You should use the same type of regulator and flowmeter for gas metal-arc welding that you use for gas tungsten-arc welding. The gas flow rates vary, depending on the types and thicknesses of the material and the joint design. At times it is necessary to connect two or more gas cylinders (manifold) together to maintain higher gas flow.

For most welding conditions, the gas flow rate is approximately 35 cubic feet per hour (cfh). This flow rate may be increased or decreased, depending upon the particular welding application. Final adjustments usually are made on a trial-and-error basis. The proper amount of gas shielding results in a rapid crackling or sizzling arc sound. Inadequate gas shielding produces a popping arc sound and results in weld discoloration, porosity, and spatter.

Regulator and flowmeter.

Regulators and flowmeters are designated for use with specific shielding gases and should be used only with the gas for which they were designed.

The hoses are normally connected to solenoid valves on the wire feeder to turn the gas flow on and off with the welding current. A hose is used to connect the flowmeter to the welding gun. The hose is often part of the welding gun assembly.

Welding Cables

Welding cables, normally made of copper or aluminium, and connectors connect the power source to the electrode holder and to the work. They consist of hundreds of wires enclosed in an insulated casing of natural or synthetic rubber. The cable that connects the power source to the welding gun is called the electrode lead. In semiautomatic welding, this cable is often part of the cable assembly, which also includes the shielding gas hose and the conduit through which the electrode wire is fed. For machine or automatic welding, the electrode lead is normally separate. The cable that connects the work to the power source is called the work lead; it is usually connected to the work by a pincer clamp or a bolt.

Recommended cable sizes for different currents and cable lengths.

Weld Type	Weld Current	Length of Cable Circuit in Feet – Cable Size AWG.					
		60'	100'	150'	200'	300'	400'
Manual (Low Duty Cycle)	100	4	4	4	2	1	1/0
	150	2	2	2	1	2/0	3/0
	200	2	2	1	1/0	3/0	4/0
	250	2	2	1/0	2/0		
	300	1	1	2/0	3/0		
	350	1/0	1/0	3/0	4/0		
	400	1/0	1/0	3/0			
	450	2/0	2/0	4/0			
	500	2/0	2/0	4/0			

Automat-	400	4/0	4/0
ic (High Duty Cycle)	800	4/0(2)	4/0(2)
	1200	4/0(3)	4/0(3)

Three factors determine the size of welding cable to use: the duty cycle of the machine, its amperage rating, and the distance between the work and the machine. If either amperage or distance increases, the cable size also must increase. Cable sizes range from the smallest at AWG No.8 to AWG No. 4/0 with amperage ratings of 75 amperes and upward. Table shows recommended cable sizes for use with different welding currents and cable lengths. A cable too small, or too long, for the current load will become too hot to handle during welding.

Other Equipment

A good ground clamp is essential to producing quality welds. Without proper grounding, the circuit voltage fails to produce enough heat for proper welding, and there is the possibility of damage to the welding machine and cables. Three basic methods are used to ground a work lead. You can fasten the ground cable to the workbench with a C-clamp, attach a springloaded clamp directly onto the workpiece, or bolt or tack-weld the end of the ground cable to the welding bench or workpiece. For a workbench, the third way creates a permanent common ground.

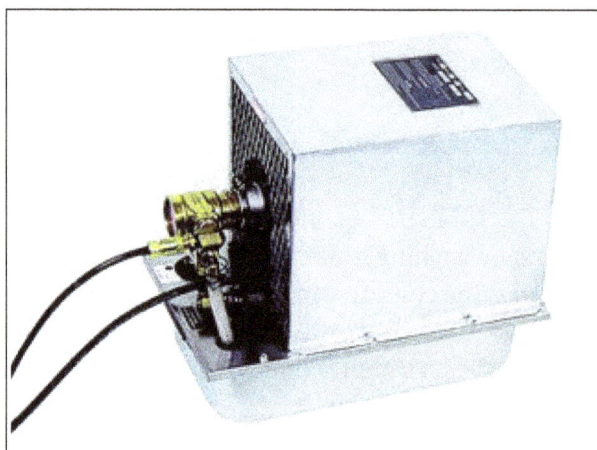

Water circulator.

Water Circulators

When a water-cooled gun is used, a water supply must be included in the system. This can be supplied by a water circulator or directly from a hose connection to a water tap. The water is carried to the welding torch through hoses that may or may not go through a valve in the welding machine. A water circulator is shown in figure.

Motion Devices

Motion devices are used for machine and automatic welding. These motion devices can be used to move the welding head, workpiece, or gun depending on the type and size of the work and the preference of the user.

Motor driven carriages that run on tracks or directly on the workpiece are commonly used. Carriages can be used for straight line contour, vertical, or horizontal welding. Side beam carriages, supported on the vertical face of a flat track, can be used for straight line welding.

Welding head manipulators may be used for longitudinal welds and, in conjunction with a rotary weld positioner, for circumferential welds. These welding head manipulators come in many boom sizes and can also be used for semiautomatic welding with mounted welding heads.

Oscillators are optional equipment used to oscillate the gun for surfacing, vertical-up welding, and other welding operations that require a wide bead. Oscillator devices can be either mechanical or electromagnetic.

Accessories

Accessory equipment used for gas metal arc welding consists of items used for cleaning the weld bead and cutting the electrode wire. In many cases cleaning is not required, but when slag is created by the welding, a chipping hammer or grinder is used to remove it. Wire brushes and grinders are sometimes used for cleaning the weld bead, and wire cutters and pliers are used to cut the end of the electrode wire between stops and starts.

Installation, Setup and Maintenance of Equipment

Learning to arc weld requires you to possess many skills. Among these skills are the abilities to set up, operate, and maintain your welding equipment.

In most factory environments, the work is brought to the welder. In the Seabees, the majority of the time the opposite is true. You will be called to the field for welding on buildings, earthmoving equipment, well drilling pipe, ship to shore fuel lines, pontoon causeways, and the list goes on. To accomplish these tasks, you have to become familiar with your equipment and be able to maintain it in the field. It would be impossible to give detailed maintenance information here because of the many different types of equipment found in the field; therefore, only the highlights will be covered.

You should become familiar with the welding machine that you will be using. Study the manufacturer's literature and check with your senior petty officer or chief on the items that you do not understand. Machine setup involves selecting current type, polarity, and current settings. The current selection depends on the size and type of electrode used, position of the weld, and the properties of the base metal.

Cable size and connections are determined by the distance required to reach the work, the size of the machine, and the amperage needed for the weld.

Operator maintenance depends on the type of welding machine used. Transformers and rectifiers require little maintenance compared to engine-driven welding machines. Transformer welders require only to be kept dry and to be given a minimal amount of cleaning. Internal maintenance should be done only by electricians due to the possibilities of electrical shock. Engine-driven machines require daily maintenance. In most places you will be required to fill out and turn in a daily inspection form called a "hard card" before starting the engine. This form is a list of items, such as oil level, water level, visible leaks, and other things, that affect the operation of the machine.

After all of these items have been checked, you are now ready to start welding.

Listed below are some additional welding rules that should be followed:

- Clear the welding area of all debris and clutter.

- Do not use gloves or clothing that contains oil or grease.

- Check that all wiring and cables are installed properly.

- Ensure that the machine is grounded and dry.

- Follow all manufacturers' directions on operating the welding machine.

- Have on hand a protective screen to protect others in the welding area from flash burns.

- Always keep fire-fighting equipment on hand.

- Clean rust, scale, paint, and dirt from the joints that are to be welded.

Power Source Connections

As a safety precaution, turn the power switches on the wire feeder and the power source to the off position before checking electrical connections. Also, always wear your safety glasses when you are in the welding area.

Check all electrical connections to make sure they are tight, and check cables for cracks and exposed wire.

On power sources that are set up for electrode positive (reverse polarity), the positive terminal that supplies welding voltage and amperage is connected to the wire feeder.

The gun trigger takes its power from a connection on the wire feeder.

The work lead is connected to the negative terminal; it should be attached to the work or to the welding table.

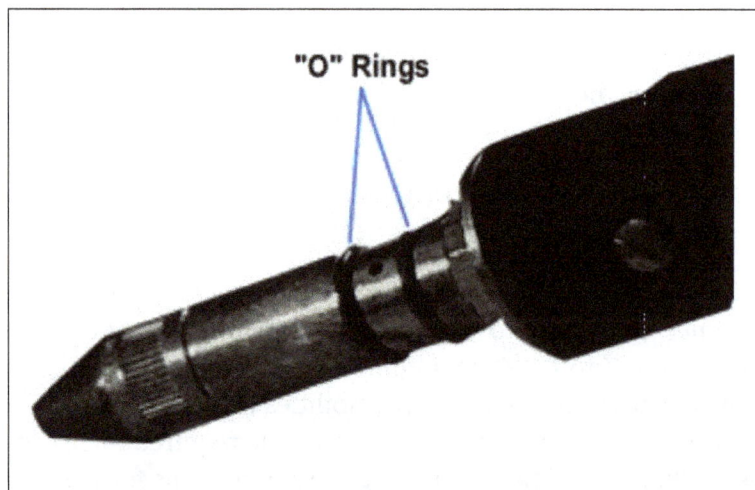

O-ring inspection.

Gun Cable Assembly

To remove the gun cable assembly: disconnect the gun trigger lead, loosen the retaining knob on the wire feeder, and pull the gun cable out of the wire feeder with a twisting motion:

- Check the O-rings for damage.

- Check the gun to make sure it is in good condition.

- Clean the nozzle.

- Use a nozzle cleaner or a pair of needle nose pliers to remove spatter from the nozzle. A dirty or damaged nozzle may interrupt the flow of shielding gas, causing porosity.

- Inspect the contact tube and gas diffuser.

- Clean spatter from the contact tube with a pair of needle nose pliers.

Replace the contact tube if the opening is worn into an oval shape. Check the gas diffuser for blockage, and clean it if necessary.

Clean the liner:

- Remove the contact tube and outlet guide.

- Stretch the cable straight.

- Blow shop air through the liner.

You should clean the liner each time you change wire to prevent dirt buildup. You should replace the liner if it is kinked or shows signs of excessive wear, such as an enlarged or oval opening. Install a new liner according to manufacturer's specification. Insert the liner into the gun cable slowly to avoid kinking it.

Wire Installation

- Remove the contact tube.

- Open the feed roll assembly. Remove the spool retaining ring.

- Slide the spool onto the spool hub so the wire feeds from bottom.

- Replace the spool retaining ring.

- Keep hand pressure on the wire to prevent the spool from uncoiling as you feed the wire through the inlet guide, across the bottom wire feed roller, and into the outlet guide.

- Close the feed roll assembly.

- Test tension by pressing the "jog" button until the wire feeds through the gas diffuser.

- Replace the contact tube and nozzle.

- Clip the wire to a 1/4 to 3/8 in. stick-out.

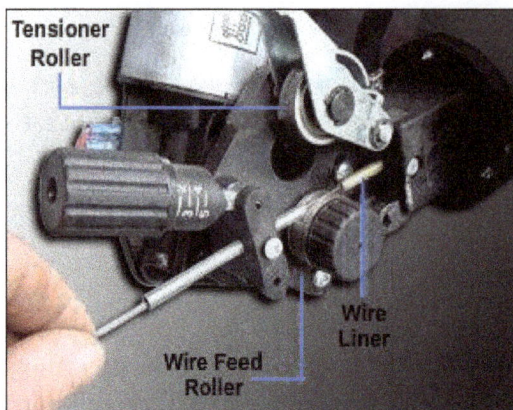
Wire Installation.

The correct amount of electrode extension or wire stick-out is important because it influences the welding current of the power source. Since the power source is self-regulating, the current output is automatically decreased when the wire stick-out increases. Conversely, when the stick-out decreases, the power source is forced to furnish more current. Too little stick-out causes the wire to fuse to the nozzle tip, which decreases the tip life.

For most GMAW, the wire stick-out should measure from 3/8 to 3/4 inch. For smaller (micro) wires, the stick-out should be between 1/4 and 3/8 inch.

Make sure the drive rolls and contact tube are matched to the diameter of the wire.

Gas Cylinder Installation

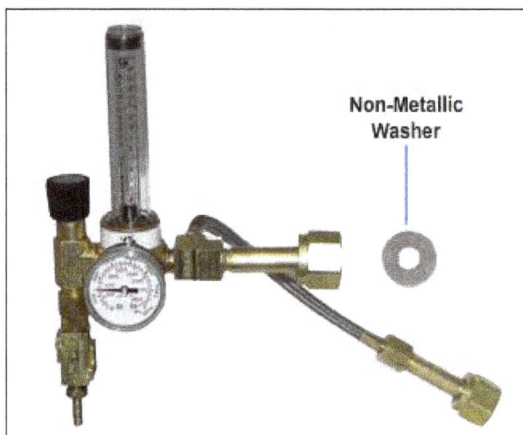
Installation of pressure regulator.

Transport a cylinder on the proper cart, chain it in place, and remove the cap.

To clear dirt from the valve opening, open and quickly close the cylinder valve.

Install the pressure regulator and flow meter assembly.

When installing 100% CO_2, insert a nonmetallic washer inside the regulator connection so the regulator does not frost. To prevent freezing for flow rates greater than 25 cubic feet per hour (cfh), use a line heater or manifold system.

Attach the gas hose to the flowmeter and wire feeder.

Open the valve slowly until pressure registers on the regulator, then open the valve completely to seat it in the fully open position.

Press the purge button and adjust the flow meter to the correct flow rate.

Amperage and Voltage Settings

Set amperage and voltage to the middle of the range specified in the welding procedure.

Fine tune the settings by performing a series of test welds.

Equipment Shutdown and Clean Up

Completely close the valve on the gas cylinder or gas manifold.

Press the purge button to bleed gas from the line.

Close the flowmeter finger tight.

Power down the wire feeder and power source.

Clean up the work area.

Burn Back

Burn back occurs when the molten tip of the electrode fuses to the end of the contact tube.

If burn back occurs, check the following:

- Voltage: If the voltage is too high in relation to the amperage, the electrode melts faster than the wire feeder can deliver wire to the puddle.

- Drive roll tension: The drive rolls could be too loose, causing the wire to slip.

- Liner and contact tube: A damaged liner or restricted contact tube may also cause burn back.

Bird Nests

Bird nests occur when the wire is impeded somewhere between the wire feeder and the work, causing the wire to pile up between the drive rolls and the outlet guide.

The most common cause of bird nests is having too much drive roll tension combined with a dirty or damaged liner, a restricted contact tube, or burnback.

To clear a bird nest:

- Clip the wire behind the inlet and outlet guides, and remove the tangle of wire.

- Remove the gun cable assembly, nozzle, and contact tube.

- Extract the wire from the back of the gun cable.

- Rethread the wire.

- Replace the contact tube and nozzle.

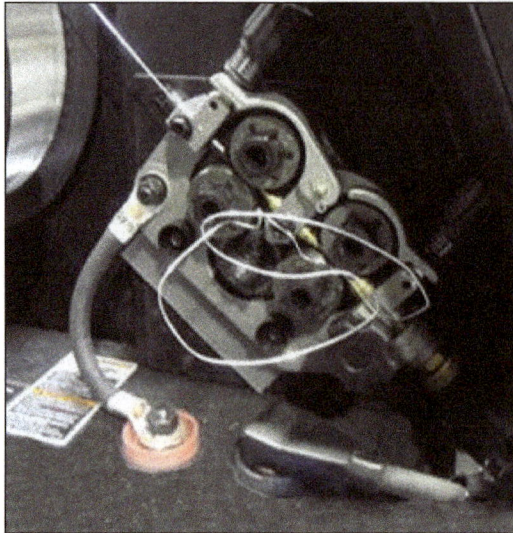

Bird nest.

Shielding Gas and Electrodes

The shielding gas is an important consumable of gas metal arc welding; its main purpose is to shield the arc and the molten weld puddle from the atmosphere. The electrodes used for this process are also consumable and provide the filler metal to the weld. The chemical composition of the electrode wire in combination with the shielding gas will determine the weld metal composition and mechanical properties of the weld.

Shielding Gases

Air in the weld zone is displaced by a shielding gas in order to prevent contamination of the molten weld puddle. This contamination is caused mainly by nitrogen, oxygen, and water vapor present in the atmosphere.

As an example, nitrogen in solidified steel reduces the ductility and impact strength of the weld and can cause cracking. In large amounts, nitrogen can also cause weld porosity.

Excess oxygen in steel combines with carbon to form carbon monoxide (CO). This gas can be trapped in the metal, causing porosity. In addition, excess oxygen can combine with other elements in steel and form compounds that produce inclusions in the weld metal.

When hydrogen, present in water vapor and oil, combines with either iron or aluminium, porosity will result, and "underbead" weld metal cracking may occur.

To avoid these problems associated with contamination of the weld puddle, three main gases are used for shielding: argon, helium, and carbon dioxide. In addition, small amounts of oxygen, nitrogen, and hydrogen have proven beneficial for some applications. Of these gases, only argon and helium are inert gases.

Both inert and active gases may be used for gas metal arc welding. When welding the non-ferrous metals, inert shielding gases are used because they do not react with the metals. The inert gases used in gas metal arc welding are argon, helium, and argonhelium mixtures.

Active or inert gases may be employed when welding the ferrous metals. Active gases such as carbon dioxide, mixtures of carbon dioxide, or oxygen-bearing shielding gases are not chemically inert and can form compounds with the metals.

Compensation for the oxidizing tendencies of other gases is made by special wire electrode formulations. Argon, helium, and carbon dioxide can be used alone, in combinations, or mixed with others to provide defect-free welds in a variety of weld applications and weld processes.

The basic properties of shielding gases that affect the performance of the welding process include the following:

- Thermal properties at elevated temperatures.

- Chemical reaction of the gas with the various elements in the base plate and welding wire.

- Effect of each gas on the mode of metal transfer.

The thermal conductivity of the gas at arc temperatures influences the arc voltage as well as the thermal energy delivered to the weld. As thermal conductivity increases, greater welding voltage is necessary to sustain the arc. For example, the thermal conductivity of helium and CO_2 is much higher than that of argon; because of this, they deliver more heat to the weld. Therefore, helium and CO_2 require more welding voltage and power to maintain a stable arc. The compatibility of each gas with the wire and base metal determines the suitability of the various gas combinations.

Carbon dioxide and most oxygen-bearing shielding gases should not be used for welding aluminium, as aluminium oxide will form. However, CO_2 and O_2 are useful at times and even essential when MIG welding steels. They promote arc stability and good fusion between the weld puddle and base material. Oxygen is a great deal more oxidizing than CO_2. Consequently, oxygen additions to argon are generally less than 12 percent by volume, whereas 100 percent CO_2 can be used for GMAW mild steels. Steel wires must contain strong deoxidizing elements to suppress porosity when used with oxidizing gases, particularly mixtures with high percentages of CO_2 or O_2 and especially 100 percent CO_2.

Shielding gases also determine the mode of metal transfer and the depth to which the workpiece is melted (depth of penetration). Summarizes recommended shielding gases for various materials and metal transfer types. Spray transfer is not obtained when the gas is rich in CO_2. For example, mixtures containing more than about 20 percent CO_2 do not exhibit true spray transfer. Rather, mixtures up to 30 percent CO_2 can have a "spray-like" shape to the arc at high current level but are unable to maintain the arc stability of lower CO mixtures. Spatter levels will also tend to increase when mixtures are rich in CO_2.

Shielding gases also determine the mode of metal transfer and the depth to which the workpiece is melted (depth of penetration). Summarizes recommended shielding gases for various materials and metal transfer types. Spray transfer is not obtained when the gas is rich in CO_2. For example, mixtures containing more than about 20 percent CO_2 do not exhibit true spray transfer. Rather, mixtures up to 30 percent CO_2 can have a "spray-like" shape to the arc at high current level but are unable to maintain the arc stability of lower CO_2 mixtures. Spatter levels will also tend to increase when mixtures are rich in CO_2.

Use of different shielding gases for gas metal arc welding.

Type of Gas	Typical Mixtures	Primary Uses
Argon		Non-ferrous metals
Helium		Aluminium, magnesium, and copper alloys
Carbon dioxide		Mild and low alloy steel
Argon-helium	20-80%	Aluminium, magnesium, copper and nickel alloys
Argon-oxygen	$1-2\%\ O_2$	Stainless steel
	$3-5\%\ O_2$	Mild and low alloy steels
Argon-carbon dioxide	$20-50\%\ CO_2$	Mild and low alloy steels
Helium-argon-carbon di-oxide	$90\%He-7\ 1/2\%Ar-2\ 1/2\%CO_2$	Stainless steel
	$60-70\%He-25-35\%Ar-5\%CO_2$	Low alloy steels
Nitrogen		Copper alloys

Several factors are usually considered in determining the type of shielding gas to be used, including the following:

- Type of metal to be welded.

- Arc characteristics and type of metal transfer.

- Speed of welding.

- Tendency to c flow meter, which interrupts the ause undercutting.

- Penetration, width, and shape of the weld bead.

- Availability.

- Cost of the gas.

- Mechanical property requirements.

Argon

Argon shielding gas is chemically inert and used primarily on the non-ferrous metals. This gas is obtained from the atmosphere by the liquification of air. Argon may be supplied as a compressed gas or a liquid, depending on the volume of use.

Argon shielding gas promotes spray type metal transfer at most current levels. Because argon is a heavier gas than helium, lower flow rates are used because the gas does not leave the welding area as fast as it does with helium. Another advantage of argon is that it gives better resistance to drafts. For any given arc length and welding current, the arc voltage is less when using argon than when using helium or carbon dioxide. This means that there is less arc energy, which makes argon preferable for welding thin metal and for metals with poor thermal conductivity.

Argon is less expensive than helium and has greater availability. It also gives easier arc starting, quieter and smoother arc action, and good cleaning action.

Helium

Helium shielding gas is chemically inert and is used primarily on aluminium, magnesium, and copper alloys. Helium is a light gas obtained by separation from natural gas. It may be distributed as a liquid but it is more often used as compressed gas in cylinders.

Helium shielding gas is lighter than air and because of this, high gas flow rates must be used to maintain adequate shielding. Typically, the gas flow rate is 2 to 3 times of that used for argon when welding in the flat position. Helium is often preferred in the overhead position because the gas floats up and maintains good shielding, while argon tends to float down. Globular metal transfer is usually obtained with helium, but spray transfer may be obtained at the highest current levels. Because of this, more spatter and a poorer weld bead appearance will be produced, as compared to argon. For any given arc length and current level, helium will produce a hotter arc, which makes helium good for welding thick metal and metals like copper, aluminium, and magnesium, which have a high thermal conductivity. Helium generally gives wider weld beads and better penetration than argon.

Carbon Dioxide

Carbon dioxide is manufactured from fuel gases given off by the burning of natural gas, fuel oil, or coke. It is also obtained as a by-product of calcination operation in lime kilns, from the manufacturing of ammonia, and from the fermentation of alcohol. The carbon dioxide given off by manufacturing ammonia and the fermenting alcohol is almost 100% pure. Carbon dioxide is made available to the user in either cylinder or bulk containers with the cylinder being more common. With the bulk system, carbon dioxide is usually drawn off as a liquid and heated to the gas state before going to the welding torch. The bulk system is normally used only when supplying a large number of welding stations.

In the cylinder, the carbon dioxide is in both a liquid and a vapor form, with the liquid carbon dioxide occupying approximately two thirds of the space in the cylinder. By weight, this is approximately 90% of the content of the cylinder. Above the liquid it exists as a vapor gas. As carbon dioxide vapor is drawn from the cylinder, it is replaced with carbon dioxide that vaporizes from the liquid in the cylinder, and therefore the overall pressure will be indicated by the pressure gage.

When the pressure in the cylinder has dropped to 200 psi (1.4 MPa), the cylinder should be replaced with a new cylinder. A positive pressure should always be left in the cylinder in order to prevent moisture and other contaminants from backing up into the cylinder. The normal discharge rate of the CO_2 cylinder is from about 4 to 35 cubic feet per hour (1.9 to 17 liters per minute). However, a maximum discharge rate of 25 cfh (12 l/min) is recommended when using a single cylinder for welding.

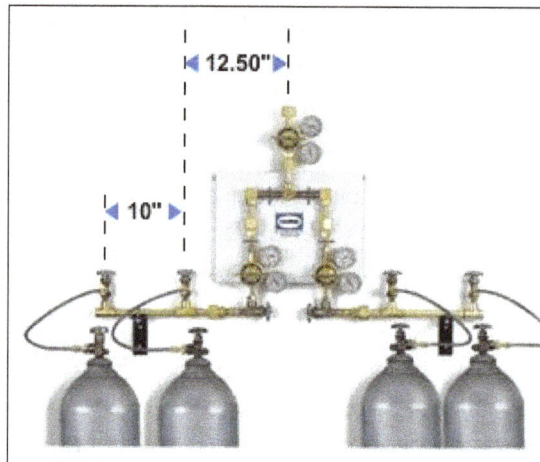

Manifold system for carbon dioxide.

As the vapor pressure drops from the cylinder pressure to discharge pressure through the CO_2 regulator, it absorbs a great deal of heat. If flow rates are set too high, this absorption of heat can lead to freezing of the regulator and flow meter, which interrupts the gas shielding. When flow rates higher than 25 cfh (12 l/min) are required, normal practice is to manifold two CO_2 cylinders in parallel or to place a heater between the bottle and gas regulator, pressure regulator, and flow-meter. A manifold system used for connecting several cylinders together. Excessive flow rates can also result in drawing liquid from the cylinder.

Carbon dioxide has become widely used for welding mild and low alloy steels. Most active gases cannot be used as shielding, but carbon dioxide offers several advantages for use in welding steel:

- Better joint penetration.

- Higher welding speeds.

- Lower welding costs (the major advantage).

Carbon dioxide produces short-circuiting transfer at low current levels and globular transfer at the higher current levels. Because carbon dioxide is an oxidizing gas, most electrode wires available for welding steel contain deoxidizers to prevent porosity in the weld. The surface of the weld bead is usually slightly oxidized even when there is no porosity.

The major disadvantage of carbon dioxide is that it produces a harsh arc and higher amounts of spatter. A short arc length is usually desirable to keep the amount of spatter to a minimum. Another problem with carbon dioxide is that it adds some carbon to the weld deposit. This does not affect mild steels, but it tends to reduce the corrosion resistance of stainless steel and reduce the ductility and toughness of the weld deposit in some of the low alloy steels.

Argon-Helium Mixtures

Regardless of the percentage, argon-helium mixtures are used for non-ferrous materials such as aluminium, copper, nickel alloys, and reactive metals. These gases used in various combinations increase the voltage and heat of GTAW and GMAW arcs while maintaining the favorable characteristics of argon. Generally, the heavier the material the higher the percentage of helium you would use. Small percentages of helium, as low as 10%, will affect the arc and the mechanical properties of the weld. As helium percentages increase, the arc voltage, spatter, and penetration will increase while minimizing porosity. A pure helium gas will broaden the penetration and bead, but depth of penetration could suffer. However, arc stability also increases. The argon percentage must be at least 20% when mixed with helium to produce and maintain a stable spray arc.

Argon-25% He (HE-25) – This little used mixture is sometimes recommended for welding aluminium where an increase in penetration is sought and bead appearance is of primary importance.

Argon-75% He (HE-75) – This commonly used mixture is widely employed for mechanized welding of aluminium greater than one inch thick in the flat position. HE-75 also increases the heat input and reduces porosity of welds in ¼-and 1/½-in. thick conductivity copper.

Argon-90% He (HE-90) – This mixture is used for welding copper over ½ in. thick and aluminium over 3 in. thick. It has an increased heat input, which improves weld coalescence and provides good X-ray quality. It is also used for short circuiting transfer with high nickel filler metals.

Argon-Oxygen Mixtures

Argon-oxygen gas mixtures usually contain 1%, 2% or 5% oxygen. The small amount of oxygen in the gas causes the gas to become slightly oxidizing, so the filler metal used must contain deoxidizers to help remove oxygen from the weld puddle and prevent porosity. Pure argon does not always provide the best arc characteristics when welding ferrous metals. In pure argon shielding, the filler metal has a tendency not to flow out to the fusion line.

The addition of small amounts of O_2 to argon greatly stabilizes the weld arc, increases the filler metal droplet rate, lowers the spray arc transition current, and improves wetting and bead shape. The weld puddle is more fluid and stays molten longer, allowing the metal to flow out towards the toe of the weld. This reduces undercutting and helps flatten the weld bead. Occasionally, small oxygen additions are used on non-ferrous applications. For example, it has been reported by NASA that .1% oxygen has been useful for arc stabilization when welding very clean aluminium plate.

- Argon-1% O_2: This mixture is primarily used for spray transfer on stainless steels. One percent oxygen is usually sufficient to stabilize the arc, improve the droplet rate, provide coalescence, and improve appearance.

- Argon-2% O_2: This mixture is used for spray arc welding on carbon steels, low alloy steels and stainless steels. It provides additional wetting action over the 1% O_2 mixture. Mechanical properties and corrosion resistance of welds made in the 1 and 2% O2 additions are equivalent.

- Argon-5% O_2: This mixture provides a more fluid but controllable weld pool. It is the most commonly used argon-oxygen mixture for general carbon steel welding. The additional oxygen also permits higher travel speeds.

- Argon-8-12% O_2: Originally popularized in Germany, this mixture has recently surfaced in the U.S. in both the 8% and 12% types. The main application is single pass welds, but some multi-pass applications have been reported. The higher oxidizing potential of these gases must be taken into consideration with respect to the wire alloy chemistry. In some instances a higher alloyed wire will be necessary to compensate for the reactive nature of the shielding gas. The higher puddle fluidity and lower spray arc transition current of these mixtures could have some advantage on some weld applications.

- Argon-12-25% O_2: Mixtures with very high O_2 levels have been used on a limited basis, but the benefits of 25% O_2 versus 12% O_2 are debatable. Extreme puddle fluidity is characteristic of this gas. A heavy slag/scale layer over the bead surface can be expected, which is difficult to remove. With care and a deoxidizing filler metal, sound welds can be made at the 25% O_2 level with little or no porosity. Removal of the slag/scale before subsequent weld passes is recommended to ensure the best weld integrity.

Argon-Carbon Dioxide Mixtures

The argon-carbon dioxide mixtures are mainly used on carbon and low alloy steels with limited application on stainless steels. The argon additions to CO_2 decrease the spatter levels usually experienced with pure CO_2 mixtures. Small CO_2 additions to argon produce the same spray arc characteristics as small O_2 additions. The difference lies mostly in the higher spray arc transition currents of argon-CO_2 mixtures. In GMAW welding with CO_2 additions, a slightly higher current level must be reached in order to establish and maintain stable spray transfer of metal across the arc. Oxygen additions reduce the spray transfer transition current. Above approximately 20% CO_2, spray transfer becomes unstable, and random short circuiting and globular transfer occur.

- Argon-3-10% CO_2: These mixtures are used for spray arc and short circuiting transfer on a variety of carbon steel thicknesses. Because the mixtures can successfully utilize both arc modes, this gas has gained much popularity as a versatile mixture. A 5% mixture is very commonly used for pulsed GMAW of heavy section low alloy steels being welding out-of-position. The welds are generally less oxidizing than those with 98 Ar-2% O_2. Improved penetration is achieved with less porosity when using CO_2 additions as opposed to O_2 additions. In the case of bead wetting, it requires about twice as much CO_2 to achieve the same wetting action as identical amounts of O_2. From 5 to 10% CO_2 the arc column becomes very stiff and defined. The strong arc forces that develop give these mixtures more tolerance to mill scale and a very controllable puddle.

- Argon-11-20% CO_2: This mixture range has been used for various narrow gap, out-ofposition sheet metal and high speed GMAW applications. Most applications are on carbon and low alloy steels. By mixing the CO_2 within this range, maximum productivity on thin gauge materials can be achieved. This is done by minimizing burn through potential while at the same time maximizing deposition rates and travel speeds. The lower CO_2 percentages also improve deposition efficiency by lowering spatter loss.

- Argon-21-25% CO_2: Used almost exclusively with short circuiting transfer on mild steel, it was originally formulated to maximize the short circuit frequency on .030- and .035- in. diameter solid wires, but through the years it has become the de facto standard for most diameter solid wire welding and has been commonly used with flux cored wires. This mixture also operates well in high current applications on heavy materials and can achieve good arc stability, puddle control, and bead appearance as well as high productivity.

- Argon-50% CO_2: This mixture is used where high heat input and deep penetration are needed. Recommended material thicknesses are above 11/8 in, and welds can be made out-of-position. This mixture is very popular for pipe welding using the short circuiting transfer. Good wetting and bead shape without excessive puddle fluidity are the main advantages for the pipe welding application. Welding on thin gauge materials has more of a tendency to burn through, which can limit the overall versatility of this gas. In welding at high current levels, the metal transfer is more like welding in pure CO_2 than previous mixtures, but some reduction in spatter loss can be realized due to the argon addition.

- Argon-75% CO_2: A 75% CO_2 mixture is sometimes used on heavy wall pipe and is the optimum in good side-wall fusion and deep penetration. The argon constituent aids in arc stabilization and reduced spatter.

Helium-Argon-Carbon Dioxide Mixtures

Three-part shielding gas blends continue to be popular for carbon steel, stainless steel, and, in restricted cases, nickel alloys. For short-circuiting transfer on carbon steel, the addition of 40% helium to argon and CO_2 as a third component to the shielding gas blend provides a broader penetration profile.

Helium provides greater thermal conductivity for short-circuiting transfer applications on carbon steel and stainless steel base materials. The broader penetration profile and increased sidewall fusion reduces the tendency for incomplete fusion.

For stainless steel applications, three-part mixes are quite common. Helium additions of 55% to 90% are added to argon and 2.5% CO_2 for short-circuiting transfer. They are favored for reducing spatter, improving puddle fluidity, and providing a flatter weld bead shape.

Common Ternary Gas Shielding Blends

- 90% Helium + 7.5% Argon + 2.5% CO_2: This is the most popular of the shortcircuiting blends for stainless steel applications. The high thermal conductivity of helium provides a flat bead shape and excellent fusion. This blend has also been adapted for use in pulsed spray transfer applications, but it is limited to stainless or nickel base materials greater than .062–in. (1.6 mm) thick. It is associated with high travel speeds on stainless steel applications.

- 55% Helium + 42.5% Argon + 2.5% CO_2: Although less popular than the 90% helium mix discussed above, this blend features a cooler arc for pulsed spray transfer. It also lends itself very well to the short-circuiting mode of metal transfer for stainless and nickel alloy applications. The lower helium concentration permits its use with axial spray transfer.

- 38% Helium + 65% Argon + 7% CO_2: This tertiary blend is for use with shortcircuiting transfer on mild and low alloy steel applications. It can also be used on pipe for open root welding. The high thermal conductivity broadens the penetration profile and reduces the tendency to cold lap.

Nitrogen

Nitrogen is occasionally used as a shielding gas when welding copper and copper alloys. Nitrogen has characteristics similar to helium because it gives better penetration than argon and tends to promote globular metal transfer. Nitrogen is used where the availability of helium is limited, such as in Europe. It can be mixed with argon for welding aluminium alloys.

Shielding Gas Flow Rate

The shielding gas flow rate should be high enough to maintain adequate shielding for the arc and weld puddle but should not be so high that it causes turbulence in the weld puddle. The gas flow rate is primarily dependent on the type of shielding gas, position of welding, and amount of electrode extension or stick-out. Higher flow rates are required for helium than for carbon dioxide and argon. These are often twice those used for carbon dioxide and argon because helium is a very light gas that floats away from the weld puddle quicker than the heavier carbon dioxide and argon gases.

In welding in the overhead position, slightly higher flow rates are often used with the heavier shielding gases because they tend to fall away from the weld puddle. The last item that affects the gas flow rate is the amount of electrode extension used. For a long electrode extension, higher gas flow rates are required to provide adequate shielding because of the greater distance between the tip of the nozzle and the weld puddle.

Electrodes

One of the most important factors to consider in GMAW welding is the correct filler wire selection. The electrode used in gas metal arc welding is bare, solid, consumable wire. In many cases, the electrode wires are chosen to match the chemical composition of the base metal as closely as possible. In some cases, electrodes with a somewhat different chemical composition will be used to obtain maximum mechanical properties or better weldability. Almost all electrodes used for gas metal arc welding of steels have deoxidizing or other scavenging elements added to minimize the amount of porosity and improve the mechanical properties. The use of electrode wires with the right amount of deoxidizers is most important when using oxygen- or carbon dioxide-bearing shielding gases.

The filler wire, in combination with the shielding gas, will produce the deposit chemistry that determines the resulting physical and mechanical properties of the weld. Five major factors influence the choice of filler wire for GMAW welding:

- Base plate chemical composition
- Base plate mechanical properties
- Shielding gas employed
- Type of service or applicable specification requirements
- Type of weld joint design

However, long experience in the welding industry has generated American Welding Society Standards to greatly simplify the selection. Wires have been developed and manufactured that consistently produce the best results with specific plate materials. Although there is no industry-wide specification, most wires conform to an AWS standard.

AWS Specification	Metal
A5.7	Copper and copper alloys
A5.9	Stainless steel
A5.10	Aluminium and aluminium alloys
A5.14	Nickel and nickel alloys
A5.16	Titanium and titanium alloys
A5.18	Carbon steel
A5.19	Magnesium alloys
A5.24	Zirconium and zirconium alloys
A5.28	Low alloy steel

Sizing

The electrodes used for gas metal arc welding are generally small in diameter when compared to the other arc welding processes. Wire diameters ranging from .030 to 1/16 in. (.8-1.6mm) are the used most widely. Wire diameters as small as .020 in. (.5mm) and up to 1/8 in. (3.2mm) are sometimes used. The electrodes are provided in a long, continuous strand of wire which is normally packaged in a coil or spool. Spools of wire normally range in weight from 2 to 60 Ibs. (.9-27 kg) and coils normally weigh 60 Ibs. (27 kg).

The electrodes' melting rates normally range from about 100 to 600 in./min. (40-255 mm/s) due to the small electrode wire sizes and the relatively high welding current levels used. Because of the small size of the electrode wire, which gives it a high surface to volume ratio, cleanliness of the wire is very important. Drawing compounds, rust, oil, or other foreign matter on the surface of the electrode wire tends to be in high proportion relative to the amount of metal present, and these items can cause weld metal defects such as porosity and cracking.

Electrode Selection

The type of metal being welded and the specific chemical and mechanical properties desired are the major factors in determining the choice of a filler metal. Identification of the base metal is absolutely required to select the proper filler metal. If the type of base metal is not known, tests can be made based on appearance, weight, magnetic check, chisel tests, flame tests, fracture tests, spark tests, and chemistry tests.

The selection of the proper filler metal for a specific job application is quite involved but can be based on the following factors:.

1. Base Metal Strength Properties - This is done by choosing a filler metal to match the tensile or yield strength of the base metal. This is usually the most important factor with low carbon and low alloy steels, as well as with some aluminium and magnesium welding applications.

2. Base Metal Chemical Compositions - The chemical composition of the base metal should be known. Closely matching the filler metal composition to the base metal composition is needed

when corrosion resistance, color match, creep resistance, and electrical or thermal conductivity are important considerations. The filler metal for non-ferrous metals, stainless steels, and many alloy steels are chosen by matching the chemical compositions.

3. Thickness and Shape of Base Metal Weldments - The workpiece may include thick sections or complex shapes, which may require maximum ductility to prevent weld cracking. Filler metal that gives the best ductility should be used.

4. Service Conditions and/or Specifications - When weldments are subjected to severe service conditions such as low temperatures, high temperatures, or shock, a filler metal that closely matches the base metal composition, ductility, and impact resistance properties should be used.

Conformances and Approvals

The electrodes used for gas metal arc welding must conform to the specifications or be approved by code-making organizations for many applications of the process. Some of the code-making organizations that issue specifications or approvals are the American Welding Society (AWS), American Society of Mechanical Engineers (ASME), American Bureau of Shipping (ABS), Federal Bureau of Roads, U.S. Coast Guard, and the Military. The American Welding Society (AWS) provides specifications for bare solid wire electrodes. The electrodes manufactured must meet specific requirements in order to conform to a specific electrode classification. Many code-making organizations such as the American Society of Mechanical Engineers (ASME) and the American Petroleum Institute (API) recognize and use the AWS specifications. Some of the code-making organizations such as the American Bureau of Shipping (ABS) and the Military must directly approve the electrodes before they can be used for welding on a project that is covered by that code. These organizations send inspectors to witness the welding and testing and to approve the classification of the solid wire electrodes.

To conform to the AWS specifications for low carbon and low alloy filler metals, the electrodes must produce a weld deposit that meets specific mechanical and chemical requirements. For the non-ferrous and stainless steel filler metal, the electrodes must produce a weld deposit with a specific chemical composition. The requirements will vary depending on the class of the electrode.

Welding Applications

Gas metal arc welding is very adaptable to many different applications. It provides the ability to weld thick metals and allows you to take your welding machine to remote locations. As you will see GMAW has become a very accepted method of welding in all industries.

Industries

Gas metal arc welding is becoming more popular for many different welding applications. When this process is used semi-automatically, higher deposition and production rates can be obtained than with the manual arc welding processes. This process is also versatile because it can be used to weld ferrous and most non-ferrous metals in all positions. It is often the only welding process practical for welding thick sections in non-ferrous metals. Gas metal arc welding lends itself easily to machine and automatic welding which are often used for producing consistent, high quality welds at the fastest travel speeds possible. This process is used extensively in the automotive industry where high production rates are required, but it is also used in the field because the equipment is relatively light and portable compared to the other continuous electrode wire processes. For this reason, gas metal arc welding is widely used in field welding of cross-country transmission pipelines and for many construction and maintenance applications.

Pressure Vessels

Gas metal arc welding is one of the more commonly used processes for welding on pressure vessels. It is used in the manufacture of plain carbon, low alloy, and stainless steel vessels as well as non-ferrous vessels. Low heat input is important on pressure vessels. Multi-layer welds are generally built up in relatively thin layers which produce better ductility and impact resistance than larger welds. Gas metal arc welding has several advantages because it produces small weld beads at much faster travel speeds than shielded metal arc welding. It also has some advantages over submerged arc welding because it can be used in all positions and the arc is not hidden beneath a flux layer. The short-circuiting and pulsed arc modes are used for out-of-position welding to reduce the heat input. The figure shows gas metal arc welding being used to weld a large mild steel vessel for an industrial refrigeration system. This process is often used for welding all passes, but sometimes it is used for welding the root passes only. Submerged arc welding is then employed for making the fill and cover passes.

GMAW Pressure Vessel Welding.

GMAW Root Pass Weld.

Industrial Piping

Gas metal arc welding also has application in the industrial piping industry. This process is widely used for welding of carbon steel, stainless steel, aluminium, copper, and nickel piping. The main advantage over shielded metal arc welding is the higher deposition rates obtained. Small diameter electrode wires are the most popular, and the shortcircuiting mode of metal transfer is widely employed. Tack welds must be carefully prepared because inadequate penetration can occur if proper variables and techniques are not used. For critical applications, skilled welders and close attention to details are required to produce complete fusion, especially on heavy parts. Thin weld layers should be avoided for this type of welding. Carbon dioxide and argon-carbon dioxide gas mixtures are used as shielding on carbon steel pipe. Open root joints in the pipe are welded in the vertical-down position when the pipe is horizontal. The rest of the weld passes may be welded either vertical up or vertical down. Gas metal arc welding is widely used for welding the fill and cover passes over a gas tungsten arc welded root pass because higher deposition rates are obtained as compared to gas tungsten arc welding.

Transmission Pipelines

Gas metal arc welding is widely used in the cross-country transmission pipeline welding industry. Most gas metal arc pipe welding is done in the field using gasoline or diesel engine driven generator-welding machines. Small diameter electrode wires are commonly employed because there is much out-of-position welding. Almost all pipes for transmission pipelines are made of carbon steel, so carbon dioxide and argon-carbon dioxide mixtures are the most popular.

Gas metal arc welding is employed using various procedures. When the process is used, most joints are welded completely with gas metal arc welding. However, some root passes are welded with shielded metal arc welding and then the joint is filled out with gas metal arc welding. A less common procedure is to use gas metal arc welding for the root pass and shielded metal arc welding for the fill and cover passes. Figure shows a root pass being welded in a 48 in. (1.2 mm) diameter natural gas pipeline.

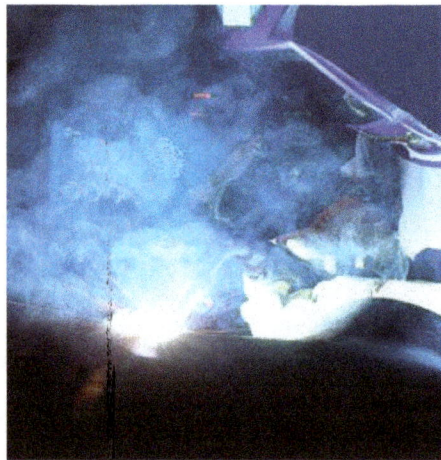

GMAW root pass of small diameter pipe.

Because the welding is being done in the field, the wind can often deflect the flow of shielding gas away from the arc. This can be prevented by setting up wind shields. An automatic welding system is sometimes employed to improve the consistency and deposition rate of the process. This equipment is normally used with special tracks that clamp on the pipe, but the equipment must be portable enough to handle in the field. When an automatic welding system is used, pipe fitup must be more precise.

Nuclear Power Facilities

Gas metal arc welding is employed but has a limited applicability in the nuclear power plants and components area. It is primarily used for welding components that are not directly part of the reactor. In the nuclear power industry, the quality of the weld deposit is the most important factor for selecting the process. Figure shows gas metal arc welding being used to weld a portion of a nuclear plenum, which is part of a nuclear filtration system. The plenum is fabricated from low carbon steel ranging in thickness from 1/16-1½ in. (1.6 -12.7 mm) and is being welded using .035 in. (.9 mm) diameter low carbon steel electrodes. Nuclear filtration systems are made of carbon or stainless steel. Other items suchas piping fittings, vessels, and liquid metal pumps are also common applications.

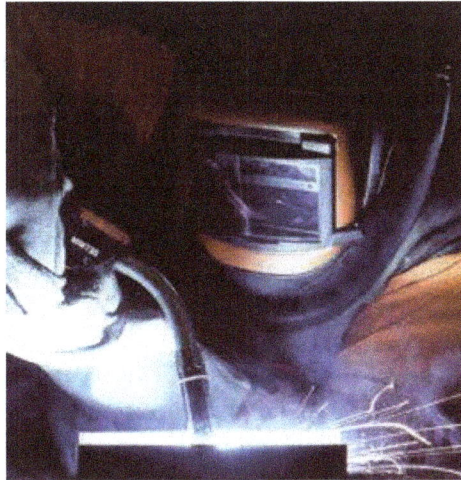

GMAW of a nuclear plenum.

Structures

The construction industry includes buildings, bridges, and other related structures. Gas metal arc welding is popular for many applications because it can be used in the field and it produces higher deposition rates than shielded metal arc welding. The development of wire feeding systems that can feed the electrode wire greater distances have helped increase the versatility of the process. The field welding applications employ gasoline or diesel engine driven generator-welding machines. The full range of electrode wire diameters is used because of the wide variety of joint designs and metal thicknesses welded.

GMAW of a structural beam.

GMAW is the most popular process for welding aluminium and other non-ferrous structures. Wind shields are often employed for field welding to prevent the loss of shielding gas. Figure shows a shop welding application where brackets are being welded on a steel structural beam. GMAW is also widely used for many multiple pass joints because of the higher deposition rates obtained.

Ships

Most of the arc welding processes are used in the shipyards, and GMAW has become widespread because of its versatility. Most ships are made of carbon steel, but nonferrous ships are welded also. Gas metal arc welding is popular because it yields higher deposition rates than shielded metal arc welding and lends itself better to welding in all positions than the other continuous wire processes.

In shipbuilding, deposition rate is the most important consideration, and because of the vast amount of welding done on a ship, GMAW is the best process for welding nonferrous metal ships and components.

Other items commonly welded are piping in the ship, non-structural components, and components that require out-of-position welding. Wire feeding systems that allow the welder to move greater distances from the source of the electrode wire are widely used. Figure shows an example of GMAW flat position welding. Portable wire feeders are often used so welders can move from one location to another more easily.

GMAW Vertical Weld.

Using .045-in. (1.1 mm) diameter electrode wire, these welds can be produced at three times the rate of shielded metal arc welding. This is a great advantage because a large percentage of the welds made in a ship are vertical fillet welds. In ship members where distortion is a problem, this process is used to get the best deposition rates with the lowest heat input.

Railroads

Gas metal arc welding is used for welding engines and cars in the railroad industry. Rail cars are fabricated from carbon steel, stainless steel, and aluminium. Machine, semiautomatic, and automatic welding are all commonly employed. GMAW and resistance welding are almost exclusively used in the manufacture of aluminium railroad cars. This process is often employed for welding in positions other than flat and for all parts of the engines and cars. Sheet metal covers for cabs, hoods, sides, and roofs are extensively welded. Because rimmed steel is widely used, filler metals of the ER70S-3 and ER70S-6 are employed; they have high amounts of de-oxidizers in them to compensate for the rimmed condition of the steel sheet metal. It is used for many sheet metal welding applications because of the fast travel speeds, which help minimize distortion problems. This process can be used for almost all components of the engines and cars, but the primary applications of the process are on thin materials and nonferrous metals, or in locations where the higher deposition rate processes, such as flux cored and submerged arc welding, cannot be used.

Automotive

In the automobile and truck manufacturing industries, both semi-automatic and automatic gas metal arc welding are widely used. It is the major process used in this industry because of the fast travel speeds obtained. Many of these applications are on items such as frames, axle housings, wheels, and body components. This process is used to weld low carbon, low alloy, and stainless steels, as well as many aluminium parts. This process is popular for welding thin sheet metal in the short-circuiting mode because it lessens the heat input and prevents burn through. The high speeds produced by this process make it very good because of the high production rates required. All thicknesses of metal are welded.

Automotive welding.

Fully automatic welding operations are used for many applications that had formerly been done using shielded metal arc welding and submerged arc welding. Gas metal arc welding has become very popular for automatic welding because it is one of the least difficult processes to fully automate. Figure shows a subframe being welded. In this application, the part is being rotated automatically, but the welder is providing joint guidance. Carbon dioxide shielding gas and a .035 in. (.9 mm) diameter electrode are being used. Gas metal arc welding is the only arc welding process being used to weld aluminium automobile body components, truck cabs, and van bodies. Figure shows the welding of an aluminium truck transmission cross-member.

Gas metal arc spot welding has many applications in the automotive and truck industries for welding the thinner metal gages of carbon steels, stainless steels, and aluminium. This process has several advantages in this industry because accessibility to the weld joint only has to be from one side, whereas resistance spot welding must have accessibility to both sides of the joint. This process is preferred because the spot welds produced have a consistent high quality and the process requires a minimum of operator skill. Typically, semi-automatic equipment is adapted for this process.

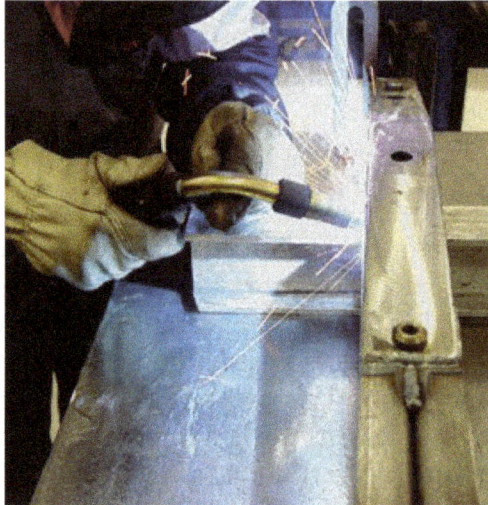

Aluminium welding.

Aerospace

GMAW is also used in the aerospace industry for many applications. It is generally employed for welding heavier sections of steel and aluminium, but it is not as widely used as gas tungsten arc welding in this industry. Gas metal arc welding allows faster travel speeds to be used, which helps minimize weld distortion and the size of the heat affected zone. Machine or automatic welding has many applications in the manufacture of in-flight refueling tanks for jet aircraft and aluminium fuel tanks for rocket motor fuel. The use of semi-automatic welding has generally been limited to less critical aircraft components. An exception to this is shown in figure where the ribbing for an aileron is being welded with a small diameter electrode wire. Gas metal arc welding is used because it can weld thin metal in all positions at high production rates.

Welding an aileron.

Heavy Equipment

Farm equipment manufacturers are major users of gas metal arc welding. It is used in the manufacture of tractors, combines, plows, tobacco harvesters, grain silos, and many other items. Other heavy equipment manufactured includes mining equipment, earthmoving equipment, and many other products. These types of equipment are generally made of mild and low carbon steels. High deposition rates are desired, so large diameter electrode wires are employed when possible. Because of this, spray and globular transfer welding are used for much of the flat position welding, but GMAW is also widely employed for producing welds in out-of-position joints.

Variations of the Process

Of the numerous variations of the GMAW process, two of the most notable are arc spot welding and narrow gap welding.

Arc Spot Welding

The gas metal arc spot welding process is used for making small localized fusion welds by penetrating through one sheet and into the other. The differences between this process and normal gas metal arc welding are that there is no movement of the welding gun and the welding takes place for only a few seconds or less. The equipment for arc spot welding usually consists of a special gun nozzle and arc timer added to a standard semi-automatic welding setup. Gas metal arc spot welding is commonly applied to mild steel, stainless steel, and aluminium, but can be used on all the metals welded by gas metal arc welding. On steel, CO_2 shielding is used to get the best penetration.

The advantages of this process over resistance spot welding are the following:

- The gun is light and portable and can be taken to the weldment.

- Spot welding can be done in all positions more easily.

- Spot welds can be made when there is accessibility only to one side of the joint.

- Spot-weld production is faster for many applications.

- Joint fitup is not as critical.

The major disadvantage of this process is that the consistency of weld strength or size is not as good as with resistance spot welding.

The weld is made by placing the welding gun on the joint. Pulling the trigger initiates the shielding gas and after a pre-flow interval, starts the arc and the wire feed. When the pre-set weld time is finished, the arc and wire feed are stopped, followed by the gas flow. The longer the weld time, the greater the penetration obtained and the higher the weld reinforcement becomes. The rest of the welding variables affect the spot weld size and shape the same way they affect a normal weld. Vertical and overhead arc spot welds can be made in metal up to .05-in. (1.3 mm) thick. For other than flat position welding, the short-circuiting mode of transfer must be used.

Many different weld joint types are made including lap, corner, and plug. The best results are obtained when the arc side member is equal to or thinner than the other. When the top plate is thicker than the bottom one, a plug weld should be made. Incomplete fusion is a common defect with this type of weld. A copper backing bar is used to prevent excessive penetration through the bottom of the weld. Another advantage of gas metal arc spot welding over resistance spot welding is that the strength can be determined from a visual examination of the weld nugget size, whereas a resistance spot weld would have to be tested to determine the strength.

Narrow Gap Welding

Narrow gap welding is another variation of the GMAW process in which square-groove or V-groove joints with small groove angles are used in thick metal sections. Root openings normally range from ¼ to 3/8 in. (6.4-9.5 mm). Narrow gap welding is generally done on ferrous metals, with the use of specially designed welding guns, but some narrow gap welding has been done on aluminium. Two small electrode wires are normally used in tandem with the wire being fed through 1 or 2 contact tubes. Each of the electrodes is fed so that the weld bead is directed toward each groove face. The special welding guns have water-cooled contact tubes and nozzles that provide shielding gas from the surface of the plate. Spray transfer is the most commonly used mode of the process, but pulsed current transfer is sometimes employed. High travel speeds are used, resulting in a low heat input and small weld puddles with narrow heat affected zones. This low heat input produces weld puddles which are easy to control in out-of position welding. Welds are normally made from one side of the plate.

Narrow gap weld.

The major problem encountered in narrow gap welding is incomplete fusion because of the low heat input in thick metal, but careful placement of the electrode wires and removing slag islands between passes to prevent slag inclusions can avoid any problems.

When used for welding metal thicknesses over 2 in. (51 mm), narrow gap welding is competitive with the other automatic arc welding processes. This type of welding has several advantages:

- Welding costs are lower because less filler metal is required.

- Lower residual stresses and less distortion are produced.

- Better welded joint properties are obtained.

The main disadvantages are the following:

- It is more prone to defects.

- Defects are more difficult to remove.

- Fitup of the joint must be more precise.

- Placement of the welding gun must be more precise.

Plasma Arc Welding

Plasma arc welding (PAW) is an arc welding process similar to gas tungsten arc welding (GTAW). The electric arc is formed between an electrode (which is usually but not always made of sintered tungsten) and the workpiece. The key difference from GTAW is that in PAW, by positioning the electrode within the body of the torch, the plasma arc can be separated from the shielding gas envelope. The plasma is then forced through a fine-bore copper nozzle which constricts the arc and the plasma exits the orifice at high velocities (approaching the speed of sound) and a temperature approaching 28,000 °C (50,000 °F) or higher.

Arc plasma is the temporary state of a gas. The gas gets ionized after passage of electric current through it and it becomes a conductor of electricity. In ionized state atoms break into electrons (−) and cations (+) and the system contains a mixture of ions, electrons and highly excited atoms. The degree of ionization may be between 1% and greater than 100% i.e.; double and triple degrees of ionization. Such states exist as more electrons are pulled from their orbits.

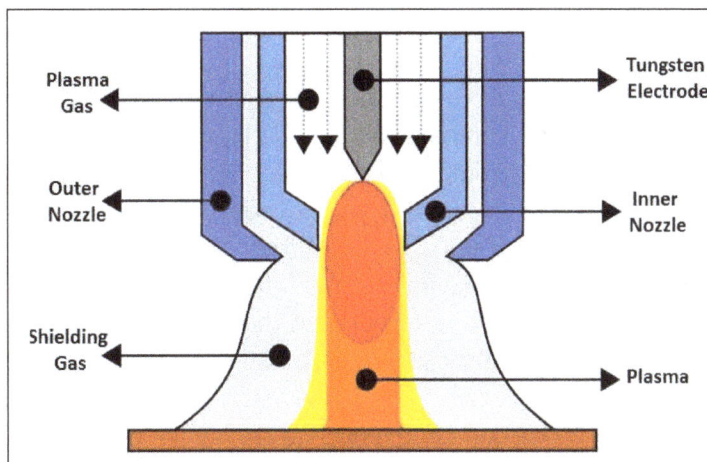

1. Gas plasma, 2. Nozzle protection, 3. Shield Gas, 4. Electrode,
5. Nozzle constriction, 6. Electric arc

The energy of the plasma jet and thus the temperature is dependent upon the electrical power employed to create arc plasma. A typical value of temperature obtained in a plasma jet torch may be of the order of 28000 °C (50000 °F) against about 5500 °C (10000 °F) in ordinary electric welding arc. Actually all welding arcs are (partially ionized) plasmas, but the one in plasma arc welding is a constricted arc plasma.

Just as oxy-fuel torches can be used for either welding or cutting, so too can plasma torches, which can achieve plasma arc welding or plasma cutting.

Plasma arc welding is an arc welding process wherein coalescence is produced by the heat obtained from a constricted arc setup between a tungsten/alloy tungsten electrode and the water-cooled (constricting) nozzle (non-transferred arc) or between a tungsten/alloy tungsten electrode and the job (transferred arc). The process employs two inert gases, one forms the arc plasma and the second shields the arc plasma. Filler metal may or may not be added.

Principle of Operation

Plasma arc welding is a constricted arc process. The arc is constricted with the help of a water-cooled small diameter nozzle which squeezes the arc, increases its pressure, temperature and heat intensely and thus improves arc stability, arc shape and heat transfer characteristics. Plasma arc welding processes can be divided into two basic types:

Non-transferred Arc Process

The arc is formed between the electrode(-) and the water cooled constricting nozzle(+). Arc plasma comes out of the nozzle as a flame. The arc is independent of the work piece and the work piece does not form a part of the electrical circuit. Just like an arc flame (as in atomic hydrogen welding), it can be moved from one place to another and can be better controlled. The non transferred plasma arc possesses comparatively less energy density as compared to a transferred arc plasma and it is employed for welding and in applications involving ceramics or metal plating (spraying). High density metal coatings can be produced by this process. A non-transferred arc is initiated by using a high frequency unit in the circuit.

Transferred Arc Process

The arc is formed between the electrode(-) and the work piece(+). In other words, arc is transferred from the electrode to the work piece. A transferred arc possesses high energy density and plasma jet velocity. For this reason it is employed to cut and melt metals. Besides carbon steels this process can cut stainless steel and nonferrous metals where an oxyacetylene torch does not succeed. Transferred arc can also be used for welding at high arc travel speeds. For initiating a transferred arc, a current limiting resistor is put in the circuit, which permits a flow of about 50 amps, between the nozzle and electrode and a pilot arc is established between the electrode and the nozzle. As the pilot arc touches the job main current starts flowing between electrode and job, thus igniting the transferred arc. The pilot arc initiating unit gets disconnected and pilot arc extinguishes as soon as the arc between the electrode and the job is started. The temperature of a constricted plasma arc may be of the order of 8000 - 25000 °C.

Equipment

The equipment needed in plasma arc welding along with their functions are as follows:

Current and Gas Decay Control: It is necessary to close the key hole properly while terminating the weld in the structure.

Fixture: It is required to avoid atmospheric contamination of the molten metal under bead.

Materials

- Steel

- Alluminium

- All the most of materials

High Frequency Generator and Current Limiting Resistors

A high frequency generator and current limiting resistors are used for arc ignition. The arc starting system may be separate or built into the system.

Plasma Torch

It is either transferred arc or non transferred arc typed. It is hand operated or mechanized. At present, almost all applications require automated system. The torch is water cooled to increase the life of the nozzle and the electrode. The size and the type of nozzle tip are selected depending upon the metal to be welded, weld shapes and desired penetration depth.

Power Supply

A direct current power source (generator or rectifier) having drooping characteristics and open circuit voltage of 70 volts or above is suitable for plasma arc welding. Rectifiers are generally preferred over DC generators. Working with helium as an inert gas needs open circuit voltage above 70 volts. This higher voltage can be obtained by series operation of two power sources; or the arc can be initiated with argon at normal open circuit voltage and then helium can be switched on.

Typical welding parameters for plasma arc welding are as follows:

Current 50 to 350 amps, voltage 27 to 31 volts, gas flow rates 2 to 40 liters/minute (lower range for orifice gas and higher range for outer shielding gas), direct current electrode negative (DCEN) is normally employed for plasma arc welding except for the welding of aluminium in which cases water cooled electrode is preferable for reverse polarity welding, i.e. direct current electrode positive (DCEP).

Shielding Gases

Two inert gases or gas mixtures are employed. The orifice gas at lower pressure and flow rate forms the plasma arc. The pressure of the orifice gas is intentionally kept low to avoid weld metal turbulence, but this low pressure is not able to provide proper shielding of the weld pool. To have suitable shielding protection same or another inert gas is sent through the outer shielding ring of the torch at comparatively higher flow rates. Most of the materials can be welded with argon, helium, argon+hydrogen and argon+helium, as inert gases or gas mixtures. Argon is very commonly used. Helium is preferred where a broad heat input pattern and flatter cover pass is desired without key hole mode weld. A mixture of argon and hydrogen supplies heat energy higher than when only argon is used and thus permits keyhole mode welds in nickel base alloys, copper base alloys and stainless steels.

For cutting purposes a mixture of argon and hydrogen (10-30%) or that of nitrogen may be used. Hydrogen, because of its dissociation into atomic form and thereafter recombination generates temperatures above those attained by using argon or helium alone. In addition, hydrogen provides a reducing atmosphere, which helps in preventing oxidation of the weld and its vicinity. (Care must be taken, as hydrogen diffusing into the metal can lead to embrittlement in some metals and steels.)

Voltage Control

Voltage control is required in contour welding. In normal key hole welding a variation in arc length up to 1.5 mm does not affect weld bead penetration or bead shape to any significant extent and thus a voltage control is not considered essential.

Process Description

Technique of work piece cleaning and filler metal addition is similar to that in TIG welding. Filler metal is added at the leading edge of the weld pool. Filler metal is not required in making root pass weld.

Type of Joints: For welding work piece up to 25 mm thick, joints like square butt, J or V are employed. Plasma welding is used to make both key hole and non-key hole types of welds.

Making a non-key hole weld: The process can make non key hole welds on work pieces having thickness 2.4 mm and under.

Making a keyhole welds: An outstanding characteristic of plasma arc welding, owing to exceptional penetrating power of plasma jet, is its ability to produce keyhole welds in work piece having thickness from 2.5 mm to 25 mm. A keyhole effect is achieved through right selection of current, nozzle orifice diameter and travel speed, which create a forceful plasma jet to penetrate completely through the work piece. Plasma jet in no case should expel the molten metal from the joint. The major advantages of keyhole technique are the ability to penetrate rapidly through relatively thick root sections and to produce a uniform under bead without mechanical backing. Also, the ratio of the depth of penetration to the width of the weld is much higher, resulting narrower weld and heat-affected zone. As the weld progresses, base metal ahead the keyhole melts, flow around the same solidifies and forms the weld bead. Key holing aids deep penetration at faster speeds and produces high quality bead. While welding thicker pieces, in laying others than root run, and using filler metal, the force of plasma jet is reduced by suitably controlling the amount of orifice gas.

Plasma arc welding is an advancement over the GTAW process. This process uses a non-consumable tungsten electrode and an arc constricted through a fine-bore copper nozzle. PAW can be used to join all metals that are weldable with GTAW (i.e., most commercial metals and alloys). Difficult-to-weld in metals by PAW include bronze, cast iron, lead and magnesium. Several basic PAW process variations are possible by varying the current, plasma gas flow rate, and the orifice diameter, including:

- Micro-plasma (< 15 Amperes)
- Melt-in mode (15–100 Amperes)
- Keyhole mode (>100 Amperes)

- Plasma arc welding has a greater energy concentration as compared to GTAW.

- A deep, narrow penetration is achievable, with a maximum depth of 12 to 18 mm (0.47 to 0.71 in) depending on the material.

- Greater arc stability allows a much longer arc length (stand-off), and much greater tolerance to arc length changes.

- PAW requires relatively expensive and complex equipment as compared to GTAW; proper torch maintenance is critical.

- Welding procedures tend to be more complex and less tolerant to variations in fit-up, etc.

- Operator skill required is slightly greater than for GTAW.

- Orifice replacement is necessary.

Process Variables

Gases

At least two separate (and possibly three) flows of gas are used in PAW:

- Plasma gas: Flows through the orifice and becomes ionized.

- Shielding gas: Flows through the outer nozzle and shields the molten weld from the atmosphere.

- Back-purge and trailing gas: Required for certain materials and applications.

These gases can all be same, or of differing composition.

Key Process Variables

- Current Type and Polarity,

- DCEN from a CC source is standard,

- AC square-wave is common on aluminium and magnesium,

- Welding current and pulsing - Current can vary from 0.5 A to 1200 A; Current can be constant or pulsed at frequencies up to 20 kHz,

- Gas flow rate (This critical variable must be carefully controlled based upon the current, orifice diameter and shape, gas mixture, and the base material and thickness).

Other Plasma Arc Processes

Depending upon the design of the torch (e.g., orifice diameter), electrode design, gas type and velocities, and the current levels, several variations of the plasma process are achievable, including:

- Plasma arc cutting (PAC).

- Plasma arc gouging.

- Plasma arc surfacing.

- Plasma arc spraying.

Plasma Arc Cutting

When used for cutting, the plasma gas flow is increased so that the deeply penetrating plasma jet cuts through the material and molten material is removed as cutting dross. PAC differs from oxy-fuel cutting in that the plasma process operates by using the arc to melt the metal whereas in the oxy-fuel process, the oxygen oxidizes the metal and the heat from the exothermic reaction melts the metal. Unlike oxy-fuel cutting, the PAC process can be applied to cutting metals which form refractory oxides such as stainless steel, cast iron, aluminium, and other non-ferrous alloys. Since PAC was introduced by Praxair Inc. at the American Welding Society show in 1954, many process refinements, gas developments, and equipment improvements have occurred.

Submerged Arc Welding

Submerged arc welding (SAW) is a common arc welding process. The first patent on the submerged-arc welding (SAW) process was taken out in 1935 and covered an electric arc beneath a bed of granulated flux. Originally developed and patented by Jones, Kennedy and Rothermund, the process requires a continuously fed consumable solid or tubular (metal cored) electrode. The molten weld and the arc zone are protected from atmospheric contamination by being "submerged" under a blanket of granular fusible flux consisting of lime, silica, manganese oxide, calcium fluoride, and other compounds. When molten, the flux becomes conductive, and provides a current path between the electrode and the work. This thick layer of flux completely covers the molten metal thus preventing spatter and sparks as well as suppressing the intense ultraviolet radiation and fumes that are a part of the shielded metal arc welding (SMAW) process.

SAW is normally operated in the automatic or mechanized mode, however, semi-automatic (handheld) SAW guns with pressurized or gravity flux feed delivery are available. The process is normally limited to the flat or horizontal-fillet welding positions (although horizontal groove position welds have been done with a special arrangement to support the flux). Deposition rates approaching 45 kg/h (100 lb/h) have been reported — this compares to ~5 kg/h (10 lb/h) (max) for shielded metal arc welding. Although currents ranging from 300 to 2000 A are commonly utilized, currents of up to 5000 A have also been used (multiple arcs).

Single or multiple (2 to 5) electrode wire variations of the process exist. SAW strip-cladding utilizes a flat strip electrode (e.g. 60 mm wide x 0.5 mm thick). DC or AC power can be used, and combinations of DC and AC are common on multiple electrode systems. Constant voltage welding power supplies are most commonly used; however, constant current systems in combination with a voltage sensing wire-feeder are available.

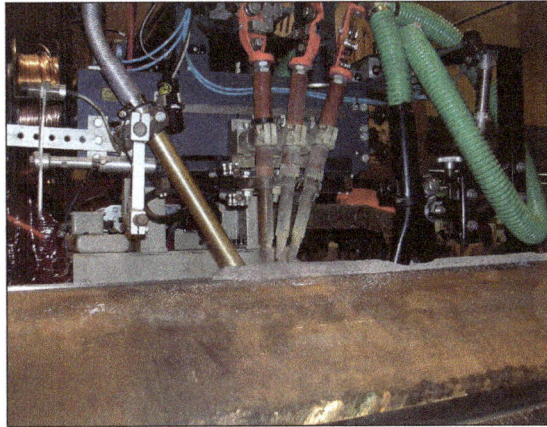

Submerged arc welding. The welding head moves from right to left.
The flux powder is supplied by the hopper on the left hand side, then follow
three filler wire guns and finally a vacuum cleaner.

A submerged arc welder used for training.

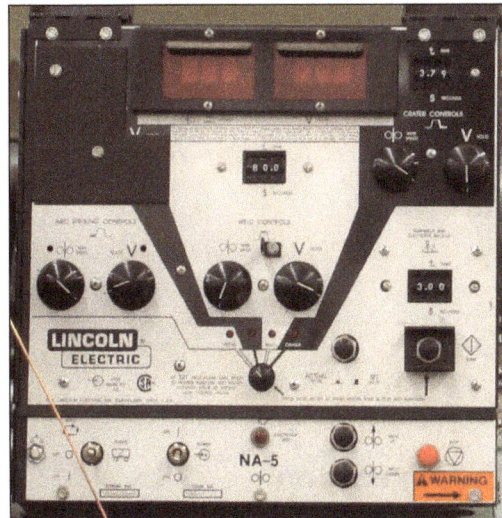

Close-up view of the control panel.

Features

- Welding head: It feeds flux and filler metal to the welding joint. Electrode (filler metal) gets energized here.

- Flux hopper: It stores the flux and controls the rate of flux deposition on the welding joint.

Flux

The granulated flux shields and thus protects molten weld from atmospheric contamination. The flux cleans weld metal and can also modify its chemical composition. The flux is granulated to a definite size. It may be of fused, bonded or mechanically mixed type. The flux may consist of fluorides of calcium and oxides of calcium, magnesium, silicon, aluminium and manganese. Alloying elements may be added as per requirements. Substances evolving large amount of gases during welding are never mixed with the flux. Flux with fine and coarse particle sizes are recommended for welding heavier and smaller thickness respectively.

Electrode

SAW filler material usually is a standard wire as well as other special forms. This wire normally has a thickness of 1.6 mm to 6 mm (1/16 in. to 1/4 in.). In certain circumstances, twisted wire can be used to give the arc an oscillating movement. This helps fuse the toe of the weld to the base metal. The electrode composition depends upon the material being welded. Alloying elements may be added in the electrodes. Electrodes are available to weld mild steels, high carbon steels, low and special alloy steels, stainless steel and some of the nonferrous of copper and nickel. Electrodes are generally copper coated to prevent rusting and to increase their electrical conductivity. Electrodes are available in straight lengths and coils. Their diameters may be 1.6, 2.0, 2.4, 3, 4.0, 4.8, and 6.4 mm. The approximate value of currents to weld with 1.6, 3.2 and 6.4 mm diameter electrodes are 150–350, 250–800 and 650–1350 Amps respectively.

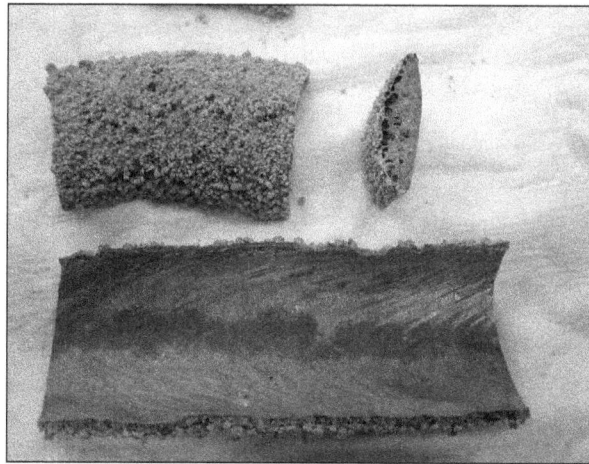

Pieces of slag from Submerged arc welding.

Welding Operation

The flux starts depositing on the joint to be welded. Since the flux when cold is non-conductor of electricity, the arc may be struck either by touching the electrode with the work piece or by placing steel wool between electrode and job before switching on the welding current or by using a high frequency unit. In all cases the arc is struck under a cover of flux. Flux otherwise is an insulator but once it melts due to heat of the arc, it becomes highly conductive and hence the current flow is maintained between the electrode and the workpiece through the molten flux. The upper portion of the flux, in contact with atmosphere, which is visible remains granular (unchanged) and can be reused. The lower, melted flux becomes slag, which is waste material and must be removed after welding.

The electrode at a predetermined speed is continuously fed to the joint to be welded. In semi-automatic welding sets the welding head is moved manually along the joint. In automatic welding a separate drive moves either the welding head over the stationary job or the job moves/rotates under the stationary welding head.

The arc length is kept constant by using the principle of a self-adjusting arc. If the arc length decreases, arc voltage will increase, arc current and therefore burn-off rate will increase thereby causing the arc to lengthen. The reverse occurs if the arc length increases more than the normal.

A backing plate of steel or copper may be used to control penetration and to support large amounts of molten metal associated with the process.

Key SAW Process Variables

- Wire feed speed (main factor in welding current control).
- Arc voltage.
- Travel speed.
- Electrode stick-out (ESO) or contact tip to work (CTTW).
- Polarity and current type (AC or DC) and variable balance AC current.

Material Applications

- Carbon steels (structural and vessel construction).
- Low alloy steels.
- Stainless steels.
- Nickel-based alloys.
- Surfacing applications (wear-facing, build-up, and corrosion resistant overlay of steels).

Advantages

- High deposition rates (over 45 kg/h (100 lb/h) have been reported).
- High operating factors in mechanized applications.
- Deep weld penetration.
- Sound welds are readily made (with good process design and control).
- High speed welding of thin sheet steels up to 5 m/min (16 ft/min) is possible.
- Minimal welding fume or arc light is emitted.
- Practically no edge preparation is necessary depending on joint configuration and required penetration.
- The process is suitable for both indoor and outdoor works.
- Welds produced are sound, uniform, ductile, corrosion resistant and have good impact value.
- Single pass welds can be made in thick plates with normal equipment.
- The arc is always covered under a blanket of flux, thus there is no chance of spatter of weld.
- 50% to 90% of the flux is recoverable, recycled and reused.

Limitations

- Limited to ferrous (steel or stainless steels) and some nickel-based alloys.
- Normally limited to the 1F, 1G, and 2F positions.
- Normally limited to long straight seams or rotated pipes or vessels.
- Requires relatively troublesome flux handling systems.
- Flux and slag residue can present a health and safety concern.
- Requires inter-pass and post weld slag removal.
- Requires backing strips for proper root penetration.
- Limited to high thickness materials.

4

Solid State Welding

Solid state welding is referred to a welding process which joins the pieces of material by using high pressure and a temperature below the melting point of the parent material. It includes explosive welding, forge welding, friction welding, diffusion welding, etc. This chapter discusses in detail these concepts of solid state welding.

Solid-state welding is a group of welding processes that produce sound joints at temperatures essentially below the melting point of the parent materials or without bulk melting of the parent materials. Solid-state welding processes have been widely applied in automobile, aircraft, and aerospace industries because of their enormous advantages associated with solid-state feature. The joints produced by solid-state processes are usually free of various solidification defects such as gas porosity, hot cracking, and nonmetallic inclusions, which may otherwise be present during fusion welding processes. No filler metals, flux, or shielding gas is required during solid-state welding process. The metal being joined can have mechanical properties similar to or even better than that of their parent metals due to the absence of defects and heat-affected zone in most of these processes. In addition, solid-state welding processes are also very suitable for joining dissimilar materials as their chemical compatibility, thermal expansion, and conductivity are no longer important problems. Solid-state welding, alternatively called solid-state bonding, covers a wide spectrum of processes including cold welding, forge welding, ultrasonic welding, friction welding, friction stir welding, resistance welding, diffusion bonding, and explosion welding. In these processes, bonding is achieved through deformation and diffusion at certain pressures and temperatures by using mechanical, electrical, or thermal energy. Unlike various fusion welding processes which are well known, solid-state welding is usually not well acquainted by industrial engineers.

Solid-State Welding Processes

Forge welding or smith welding is the oldest known welding process and its use has been reported from 1400 B.C. By this process the pieces to be welded are heated to above 1000 °C and then placed together and given impact blows by hammering. In the more recent form of large welding the pressure is applied by rolling, drawing and squeezing to achieve the forging action.

The oxides are excluded by virtue of design of the workpieces and or by the use of appropriate temperature as well as fluxes. Fluxes commonly used for forge welding low carbon steels are sand, fluorspar and borax. They help in melting the oxides, if formed.

Proper heating of the workpieces is the major welding variable that controls the joint quality. Insufficient heating may not affect a joint while overheating results in a brittle joint of low strength. Also, the overheated pieces tend to be oxidised which shows itself by spongy appearance.

An excellent living example of forge welded component of the olden days is the Iron Pillar of Delhi which measures 7-6 m in length with an average diameter of 350 mm and weighs 5.4 tonnes. These days the process is mainly used for welding low carbon steel parts usually for agriculture implements in rural areas of third world countries.

In friction welding one piece is held stationary and the other is rotated in the chuck of a friction welding machine. As they are brought to rub against each other under pressure, they get heated due to friction. When the desired forging temperature is attained throughout the rubbing cross-section of the workpieces, the rotation is stopped suddenly and the axial pressure is increased to cause a forging action and hence welding. This method has been in use for welding of thermoplastics since 1945 but metals were first welded successfully by it in 1956.

The machine used for friction welding resembles a lathe but is sturdier than that. The essential features of the machine are that it should be able to withstand high axial pressure of the order of 50,000 N/cm² and be able to provide a high spindle speed of upto 12,000 rpm though the usual range may rarely exceed 5000 rpm.

A less popular variant of the process is called INERTIA WELDING in which welding is achieved by the rotation of a flywheel which is detached at the desired moment and comes to a stop within the stipulated time, thus eliminating the braking unit. The principles of continuous drive and inertia type friction welding processes.

Friction welding is a high speed process suited to production welding. However, initial trials are required to standardise the process parameters for a given job. Friction welding of two pieces rarely takes more than 100 seconds though it may be just about 20 seconds for small components.

One of the parts to be friction welded needs to be round which puts a serious limitation on the use of this process. However, it is increasing in popularity and can weld most of the metals and their dissimilar combinations such as copper and steel, aluminium and steel, aluminium and titanium, etc. Typical applications of the process include welding of drill bits to shanks, i.e. engine valve heads to stems, automobile rear-axle hub-end to axle casing.

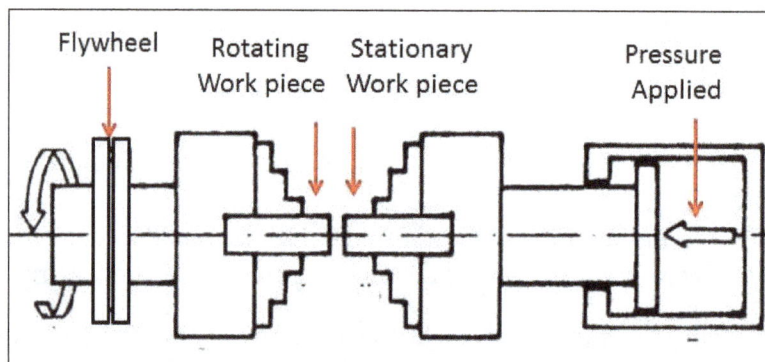

Stages in two types of friction welding.

In explosive or explosion welding process the weld is achieved by making one part strike against the other at a very high but subsonic velocity. This is achieved by the use of explosives usually of the ammonium nitrate base. The process is completed in micro-seconds.

The setup, in principle, used for explosive welding is figure shown in the two plates to be welded placed at an inclination to each other. The included angle varies between 1° and 10°. The thicker plate called the target plate is placed on an anvil and the thinner plate called the flyer plate has a buffer plate of PVC or rubber, between it and the explosive charge, for protection against surface damage.

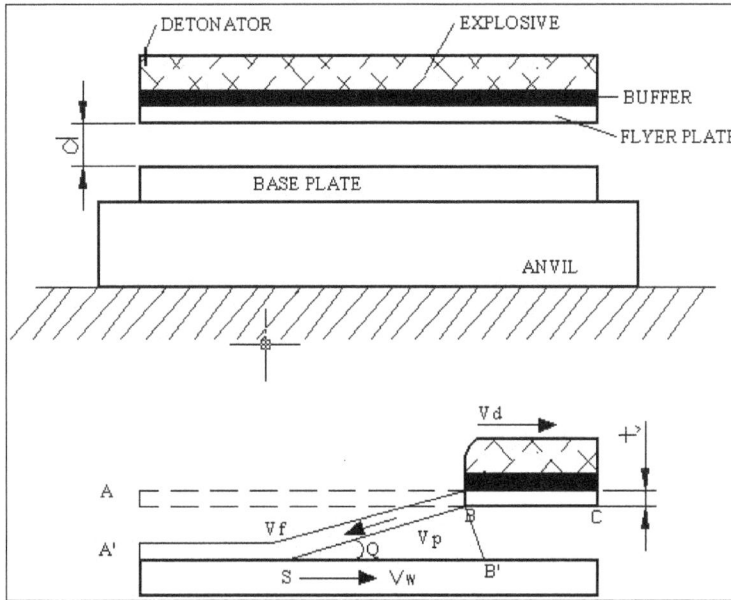

Schematic of explosion welding setup.

The charge is exploded by a detonator placed at one end of the flyer plate. When the charge explodes, the flyer plate moves towards the target plate at a velocity of 150 to 550 m/sec and the pressure produced at the interface of the impacting plates by such a high velocity is of the order of 70,000 to 700,000 N/cm².

Under such a high velocity and pressure the metal flows ahead of the joining front acting like a fluid jet resulting in a bond of the interlocking type. This interlocking is an essential aspect of an explosion weld and is the cause of its strength. The weld strength equal to the strength of the weaker of the two components (metals) can be achieved.

Explosive welding is normally an outdoor process and needs a large area to ward-off the persons coming close to the explosion site particularly when an explosive charge of high strength may have to be exploded.

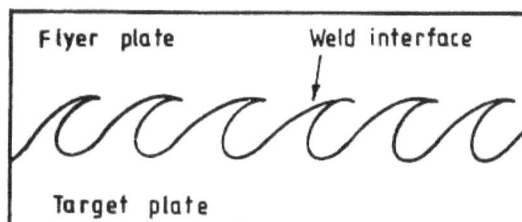

Weld interface in an explosion weld.

Explosive welding can be used for welding dissimilar metal combinations like copper and steel, aluminium and mild steel, aluminium and Inconel (76% Ni + 15% Cr + Fe), aluminium and stainless steel, etc. It can also be used for welding tantalum, titanium, and nickel components.

Typical applications of explosive welding include cladding of thick plates by thin sheets, even foils. Tube to tub-sheet joints in heat exchangers, valve to pipe joint, as well as blocking of leaking tubes in boilers can be successfully achieved by this process.

It is a pressure welding process which is employed at a temperature above 200 °C. The process deals with mainly small components in the electrical and electronic industries for welding fine wires of about 0.025 mm diameter to metal films on glass or ceramic.

There are many versions of the process, three out of which and are referred to as chisel or wedge bond, ball bond, and parallel gap bond. In the chisel or wedge bond a wire is deformed, under pressure, and welded to the film with the help of wedge shaped indentor. In the ball bond a wire is heated by a micro-hydrogen flame to form a ball at the wire tip which is subsequently welded to the heated film on substrate by the pressure exerted through the pierced indentor.

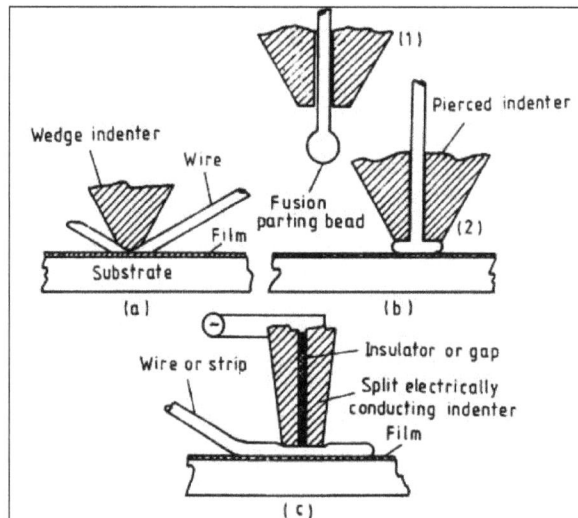

Thermo-compression bonding techniques: (a) wedge bond
(b) ball bond (c) resistance or parallel- gap bond.

In the parallel gap bond the wire or strip is pressed to the film with the help of twin electrode made of high resistance material like tungsten. The flow of current through the wire or strip heats it up locally thus keeping the heat confined to the small zone around it.

For all these variants of the process local inert atmosphere is created around the joint being bonded. Ultrasonic vibrations replace heating in some of the applications of all these modes of the process.

Commercial applications of the process include welding of noble metals, aluminium, and copper to substrates of glass, or ceramic.

In diffusion bonding or diffusion welding a weld is achieved by the application of pressure, of the order of 5 to 75 N/mm^2, while the pieces are held at a high temperature, normally about 70% of the melting point in degrees absolute i.e. about 1000 °C for steel. The process is based on solid-phase diffusion which, obviously, is accelerated with rise in temperature.

Diffusion in metals takes place due to vacant lattice sites or along grain boundaries, and is expressed by the following mathematical relationship:

$$D = D_0 e^{-ERT}$$

where,

D = rate of diffusion,

D_0 = Constant having the same dimension as D,

E = activation energy,

R = gas constant,

T = Absolute temperature at which the workpieces are held.

Depending upon the extent of diffusion required the process may be completed in 2 to 3 minutes or may take many minutes or even hours. The quality of surfaces to be welded plays an important role. A good quality surface turned, milled or ground to a standard of 0-4 to 0-2 μm* CLA (centre-line average) is usually adequate. The surface must be degreased before welding by using acetone or petroleum ether swab.

Presence of oxide layers on the surfaces being joined hinder diffusion but get dissipated over a period of time. Thus, metals which dissolve their own oxides such as iron and titanium, bond easily. On the contrary metals that form tough refractory oxide layers, like aluminium, are difficult to diffusion weld.

Diffusion Bonding can be Achieved by three Methods viz

- Gas pressure bonding,

- Vacuum fusion bonding, and

- Eutectic fusion bonding.

In gas pressure bonding, the parts are held together in an inert atmosphere and heated to a temperature of 800 °C by a system resembling an autoclave. During heating the high pressure provides uniform pressure over all the surfaces. This method is used for bonding non-ferrous metals only because it necessitates high temperatures for steels.

In vacuum fusion bonding the parts are held in an intimate contact in a vacuum chamber. The pressure on the parts is applied by mechanical means or a hydraulic pump, and heating is done in the same way as in gas pressure welding. A schematic for vacuum fusion bonding.

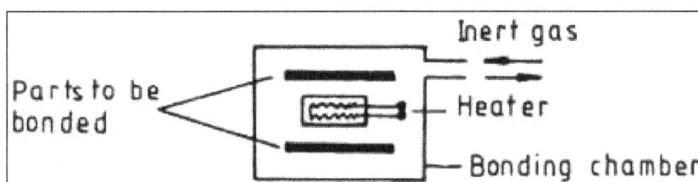

A setup for diffusion bonding.

A vacuum pumping system which can quickly reduce pressure to about 10-3 torr (mm of mercury) needs to be used. High pressure created by the use of mechanical or hydraulic means makes it possible to diffusion bond steels by this method. Successful joining of steel can be achieved at a temperature of about 1150 °C under an applied pressure of nearly 70 N/mm².

In eutectic fusion bonding a thin piece of a particular material is placed between the surfaces to be welded. This results in the formation of a eutectic compound by diffusion at an elevated temperature and the piece may completely disappear and form eutectic alloy(s) at the interface. The material used for being placed in between the two parts is usually a dissimilar metal in foil form with a thickness of 0-005 to 0-025 mm.

Diffusion bonding can be used to join dissimilar metals e.g., steel can be welded to aluminium, tungsten, titanium, molybdenum, cermet's (compounds of ceramics and metals), copper to titanium, titanium to platinum, etc. It finds use in radio engineering, electronics, instrument making, missile, aircraft, nuclear, and aerospace industries.

Typical applications of diffusion bonding include tipping of heavy cutting tools with carbide tips or hard alloys, joining of vacuum tube components, fabrication of high temperature heaters from molybdenum disilicide for resistor furnace that can operate in an oxidising atmosphere upto 1650 °C. In aerospace industry it is used for fabricating complex shaped components of titanium from simple structural shapes. It is also used for surfacing components to resist wear, heat or corrosion.

Forge Welding

Forge welding (FOW) is a solid-state welding process that joins two pieces of metal by heating them to a high temperature and then hammering them together. It may also consist of heating and forcing the metals together with presses or other means, creating enough pressure to cause plastic deformation at the weld surfaces. The process is one of the simplest methods of joining metals and has been used since ancient times. Forge welding is versatile, being able to join a host of similar and dissimilar metals. With the invention of electrical and gas welding methods during the Industrial Revolution, manual forge-welding has been largely replaced, although automated forge-welding is a common manufacturing process.

Forge welding is a process of joining metals by heating them beyond a certain threshold and forcing them together with enough pressure to cause deformation of the weld surfaces, creating a metallic bond between the atoms of the metals. The pressure required varies, depending on the temperature, strength, and hardness of the alloy. Forge welding is the oldest welding technique, and has been used since ancient times.

Welding processes can generally be grouped into two categories: fusion and diffusion welding. Fusion welding involves localized melting of the metals at the weld interfaces, and is common in electric or gas welding techniques. This requires temperatures much higher than the melting point of the metal in order to cause localized melting before the heat can thermally conduct away from the weld, and often a filler metal is used to keep the weld from segregating due to the high surface tension. Diffusion welding consists of joining the metals without melting them, welding the surfaces together while in the solid state. In diffusion welding, the heat source is often lower than the melting point of the metal, allowing more even heat-distribution thus reducing thermal stresses at the weld. In this method a filler metal is typically not used, but the weld occurs directly between the metals at the weld interface. This includes methods such as cold welding, explosion welding, and forge welding. Unlike other diffusion methods, in forge welding the metals are heated to a high temperature before forcing them together, usually resulting in greater plasticity at the weld surfaces. This generally makes forge welding more versatile than cold-diffusion techniques, which are usually performed on soft metals like copper or aluminium. In forge welding, the entire welding areas are heated evenly. Forge welding can be used for a much wider range of harder metals and alloys, like steel and titanium.

Materials

Many metals can be forge welded, with the most common being both high and low-carbon steels. Iron and even some hypoeutectic cast-irons can be forge welded. Some aluminium alloys can also be forge welded. Metals such as copper, bronze and brass do not forge weld readily. Although it is possible to forge weld copper-based alloys, it is often with great difficulty due to copper's tendency to absorb oxygen during the heating. Copper and its alloys are usually better joined with cold welding, explosion welding, or other pressure-welding techniques. With iron or steel, the presence of even small amounts of copper severely reduces the alloy's ability to forge weld.

Titanium alloys are commonly forge welded. Because of titanium's tendency to absorb oxygen when molten, the solid-state, diffusion bond of a forge weld is often stronger than a fusion weld in which the metal is liquefied.

Forge welding between similar materials is caused by solid-state diffusion. This results in a weld that consists of only the welded materials without any fillers or bridging materials. Forge welding between dissimilar materials is caused by the formation of a lower melting temperature eutectic between the materials. Due to this the weld is often stronger than the individual metals.

Processes

The most well-known and oldest forge-welding process is the manual-hammering method. Manual hammering is done by heating the metal to the proper temperature, coating with flux, overlapping the weld surfaces, and then striking the joint repeatedly with a hand-held hammer. The joint is often formed to allow space for the flux to flow out, by beveling or rounding the surfaces slightly, and hammered in a successively outward fashion to squeeze the flux out. The hammer blows are typically not as hard as those used for shaping, preventing the flux from being blasted out of the joint at the first blow.

When mechanical hammers were developed, forge welding could be accomplished by heating the metal, and then placing it between the mechanized hammer and the anvil. Originally powered by waterwheels, modern mechanical-hammers can also be operated by compressed air, electricity, steam, gas engines, and many other ways. Another method is forge welding with a die, whereas the pieces of metal are heated and then forced into a die which both provides the pressure for the weld and keeps the joint at the finished shape. Roll welding is another forge welding process, where the heated metals are overlapped and passed through rollers at high pressures to create the weld.

Modern forge-welding is often automated, using computers, machines, and sophisticated hydraulic-presses to produce a variety of products from a number of various alloys. For example, steel pipe is often forge-welded during the manufacturing process. Flat stock is heated and fed through specially-shaped rollers that both form the steel into a tube and simultaneously provide the pressure to weld the edges into a continuous seam. Diffusion bonding is a common method for forge welding titanium alloys in the aerospace industry. In this process the metal is heated while in a press or die. Beyond a specific critical-temperature, which varies depending on the alloy, the impurities burn out and the surfaces are forced together. Other methods include flash welding and percussion welding. These are resistance forge-welding techniques where the press or die is electrified, passing high current through the alloy to create the heat for the weld. Shielded active-gas forge-welding is a process of forge welding in an oxygen-reactive environment, to burn out oxides, using hydrogen gas and induction heating.

Temperature

The temperature required to forge weld is typically 50 to 90 percent of the melting temperature. Iron can be welded when it surpasses the critical temperature (the A4 temperature) where its allotrope changes from gamma iron (face-centered cubic) to delta iron (body-centered cubic). Since the critical temperatures are affected by alloying agents like carbon, steel welds at a lower temperature-range than iron. As the carbon content in the steel increases, the welding temperature-range decreases in a linear fashion. Iron, different steels, and even cast-iron can be welded to each other, provided that their carbon content is close enough that the welding ranges overlap. Pure iron can be welded when nearly white hot; between 2,500 °F (1,400 °C) and 2,700 °F (1,500 °C). Steel with a carbon content of 2.0% can be welded when orangish-yellow, between 1,700 °F (900 °C) and 2,000 °F (1,100 °C). Common steel, between 0.2 and 0.8% carbon, is typically welded at a bright yellow heat.

A primary requirement for forge welding is that both weld surfaces need to be heated to the same temperature and welded before they cool too much. When steel reaches the proper temperature, it begins to weld very readily, so a thin rod or nail heated to the same temperature will tend to stick at first contact, requiring it to be bent or twisted loose. One of the simplest ways to tell if iron or steel is hot enough is to stick a magnet to it. When iron crosses the A2 critical temperature, it begins to change into the allotrope called gamma iron. When this happens, the steel or iron becomes non-magnetic. In steel, the carbon begins to mix with gamma iron at the A3 temperature, forming a solid solution called austenite. When it crosses the A4 critical temperature, it changes into delta iron, which is magnetic. Therefore, a blacksmith can tell when the welding temperature is reached by placing a magnet in contact with the metal. When red or orange-hot, a magnet will not stick to the metal, but when the welding temperature is crossed, the magnet will again stick to it. The steel may take on a glossy or wet appearance at the welding temperature. Care must be taken to avoid overheating the metal to the point that it gives off sparks from rapid oxidation (burning), or else the weld will be poor and brittle.

Decarburization

When steel is heated to an austenizing temperature, the carbon begins to diffuse through the iron. The higher the temperature; the greater the rate of diffusion. At such high temperatures, carbon readily combines with oxygen to form carbon dioxide, so the carbon can easily diffuse out of the steel and into the surrounding air. By the end of a blacksmithing job, the steel will be of a lower carbon content than it was prior to heating. Therefore, most blacksmithing operations are done as quickly as possible to reduce decarburization, preventing the steel from becoming too soft.

To produce the right amount of hardness in the finished product, the smith generally begins with steel that has a carbon content that is higher than desired. In ancient times, forging often began with steel that had a carbon content much too high for normal use. Most ancient forge-welding began with hypereutectoid steel, containing a carbon content sometimes well above 1.0%. Hypereutectoid steels are typically too brittle to be useful in a finished product, but by the end of forging the steel typically had a high carbon-content ranging from 0.8% (eutectoid tool-steel) to 0.5% (hypoeutectoid spring-steel).

Applications

Forge welding has been used throughout its history for making most any items out of steel and iron. It has been used in everything from the manufacture of tools, farming implements, and cookware to the manufacture of fences, gates, and prison cells. In the early Industrial Revolution, it was commonly used in the manufacture of boilers and pressure vessels, until the introduction of fusion-welding. It was commonly used through the Middle Ages for producing armor and weapons.

One of the most famous applications of forge welding involves the production of pattern-welded blades. During this process a smith repeatedly draws out a billet of steel, folds it back and welds it upon itself. Another application was the manufacture of shotgun barrels. Metal wire was spooled onto a mandrel, and then forged into a barrel that was thin, uniform, and strong. In some cases the forge-welded objects are acid-etched to expose the underlying pattern of metal, which is unique to each item and provides aesthetic appeal.

Despite its diversity, forge welding had many limitations. A primary limitation was the size of objects that could be forge welded. Larger objects required a bigger heat source, and size reduced the ability to manually weld it together before it cooled too much. Welding large items like steel plate or girders was typically not possible, or at least highly impractical, until the invention of fusion welding, requiring them to be riveted instead. In some cases, fusion welding produced a much stronger weld, such as in the construction of boilers.

Flux

Forge welding requires the weld surfaces to be extremely clean or the metal will not join properly, if at all. Oxides tend to form on the surface while impurities like phosphorus and sulfur tend to migrate to the surface. Often a flux is used to keep the welding surfaces from oxidizing, which would produce a poor quality weld, and to extract other impurities from the metal. The flux mixes with the oxides that form and lowers the melting temperature and the viscosity of the oxides. This enables the oxides to flow out of the joint when the two pieces are beaten together. A simple flux can be made from borax, sometimes with the addition of powdered iron-filings.

The oldest flux used for forge welding was fine silica sand. The iron or steel would be heated in a reducing environment within the coals of the forge. Devoid of oxygen, the metal forms a layer of iron-oxide called wustite on its surface. When the metal is hot enough, but below the welding temperature, the smith sprinkles some sand onto the metal. The silicon in the sand reacts with the wustite to form fayalite, which melts just below the welding temperature. This produced a very effective flux which helped to make a strong weld.

Early examples of flux used different combinations and various amounts of iron fillings, borax, sal ammoniac, balsam of copaiba, cyanide of potash, and soda phosphate. The 1920 edition of Scientific American book of facts and formulae indicates a frequently offered trade secret as using copperas, saltpeter, common salt, black oxide of manganese, prussiate of potash, and "nice welding sand" (silicate).

Cold Pressure Welding

Cold pressure welding is a form of solid phase welding, which is unique because it is carried out at ambient temperatures. (Other forms of solid phase welding are conducted at elevated temperatures, but although these temperatures are high, the material is not molten, merely more ductile.)

As early as 3,000 BC, the Egyptians prepared iron by hammering a metal sponge in order to weld the red-hot particles together. Blacksmiths have also hammer welded wrought iron for centuries. This type of welding was always carried out at high temperatures.

The first known example in Britain of hammer welding at ambient temperatures (therefore true cold pressure welding) dates back to the late Bronze Age, around 700 BC. The material used was gold, and gold boxes made by this process have been found during excavations.

The first scientific observation of cold pressure welding was made in 1724 by the Reverend J I Desaguliers. He demonstrated the phenomenon to the Royal Society and later published the details in the scientific journals of the time. Reverend Desaguliers discovered that if he took two lead balls about 25mm each in diameter, pressed them together and twisted them, then the two pieces would join together. The joint strength was measured on a steelyard and although the results were erratic, good bonds were produced, with some as strong as the parent material.

After Reverend Desaguliers' discovery in the 18th century, it appears that very little happened until the Second World War. This accelerated developments, especially in Germany, where light alloy cooler elements for aircraft were pressure welded, although it is understood that this welding was carried out at elevated temperatures.

Seen for the first time, cold pressure welding can appear an almost magical process. People unfamiliar with it are often reluctant to accept a method of welding that does not involve heat or electricity and some form of flux to make the joins. After a demonstration, they inevitably ask, "How are the two pieces of metal joined?"

There have been several explanations as to the actual mechanism by which a cold pressure weld is obtained. For example, it has been suggested that it happens via recrystallisation or by an energy hypothesis, but most explanations have been either experimentally disproved or refuted on theoretical grounds.

The currently accepted hypothesis that accounts for a cold pressure weld taking place involves the atoms of metals being held together by the metallic 'bond', so called because it is peculiar to metallic substances. The bond can be described as a 'cloud' of free, negatively charged atoms formed into a unit as a result of attractive forces.

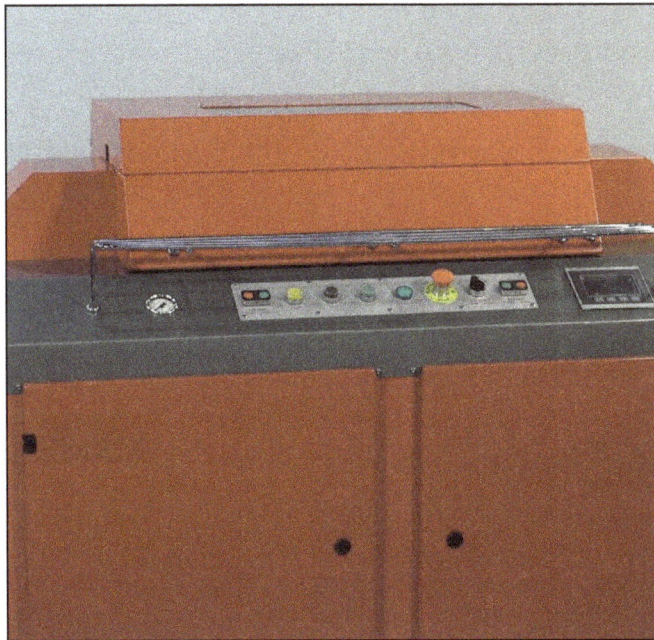

The electro/pneumatic EP500 rod welder is a heavy-duty machine that will weld non-ferrous wire and strip from 5.00mm (.197") up to 12.50mm (.492").

Creating a Weld

Therefore, if two metallic surfaces are brought together with only a few angstroms separation (there being 300 million angstroms to one centimetre) interaction between the free electrons and ionised atoms can occur. This will eliminate the potential barrier, allowing the electron cloud to become common. This, in turn, results in a bond and therefore a weld.

A simpler way of explaining this rather awesome process is that if two surfaces are put together, both being anatomically clean and anatomically flat when considered on an atomic scale, a bond is effected equal to that of the parent material.

Lengths of welded copper/aluminium rod.

Early Applications

In practice, however, bonding is virtually impossible under most conditions, because of surface irregularities, organic surface contamination and chemical films such as oxide films.

In order to obtain maximum weld efficiency, any form of contamination must be reduced to a minimum, while the area of contact, the weld area, has to be made as large as possible.

In earlier applications of cold pressure butt welding, the upset and radial displacement of the interfaces was undertaken in a single step. This technique had several disadvantages: it was necessary to square off the ends to be joined; both surfaces had to be kept free of contamination; and the amount of material which projected from the gripping die was such that bending and lack of coaxiality could occur, thereby spoiling the correct flow of metal.

Multi Upset Principle

The M10 hand-held, manually operated cold welder will join fine
wire from 0.10mm (.0039) to 0.50mm (.0196).

Then came the system of butt welding developed by GEC, employing what is known as the 'multi upset principle'. When the material is inserted in the die, each time the machine is activated, the material is gripped by the die and fed forward.

In this way, the two opposing faces are stretched and enlarged over their entire surface area as they are pushed against each other. The oxide and other surface impurities are forced outward from the core of the material and a bond is effected. A minimum of four upsets is recommended to ensure all impurities are squeezed out of the interfaces.

The advantages of this type of welding are easily seen in practice. The ends of the wire or rod need no preparation prior to welding and the alignment of the two butt ends is automatic as the material is placed in the die. There is no heat setting to be arrived at; no gap setting to be made, as this is built into the die; and no spring pressure to be set. Any one of these things incorrectly set on a resistance butt welder would result in a weld failure.

Suitable Metals

Cold pressure welding is restricted to nonferrous materials or, at best, soft iron that has no carbon content. Most nonferrous metals can be cold welded, and while copper and aluminium are the most common, various alloys such as Aldrey, Triple E, Constantan, 70/30 brass, zinc, silver and silver alloys, nickel, gold and many others have good cold weldability. Plated wires, including tinned copper, silver plated and nickel-plated, can all be welded to themselves or to plain copper.

The usual methods of joining dissimilar metals such as copper and aluminium, namely resistance welding, friction welding or flame brazing, will all result in a rapid breakdown of the joint. This reaction in a copper/aluminium joint begins to take place as soon as the two metals are placed together.

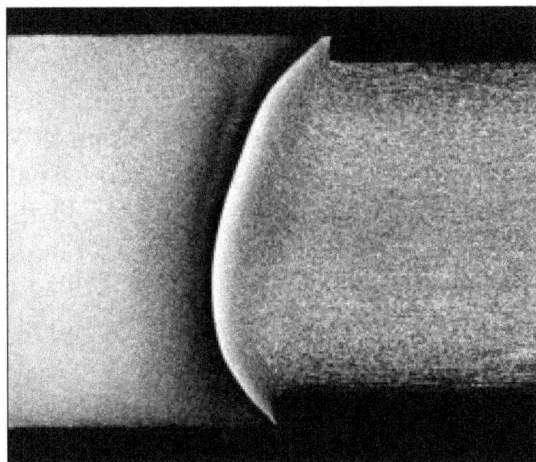

A cross section of a welded area, showing 8mm (0.315") diameter copper
rod joined to 9.5mm (0.374") diameter aluminium rod.

The problem is created by the oxides and the air space, which are left between the interfaces during these methods of welding, rather than by the dissimilarity between the metals themselves. However, with cold pressure welding, these oxides and air spaces are squeezed out in the weld process and, since no heat is applied, only the metallurgical changes that operate at ambient temperatures occur.

Cold pressure welding provides the most satisfactory way of joining copper to aluminium without the formation of brittle inter-metallic compounds. The quality is excellent because it produces a worked structure as opposed to the cast structure obtained in fusion welding. Also, there is no heat-affected zone with unsuitable properties.

To test weld strength, most people rely on a tensile tester. Alternatively, you can make a reverse bend test. However, the most stringent test is to pass the weld though a number of dies in a wire drawing machine.

The Role of Dies

The dies play an important role in the cold butt weld process. Firstly, they must grip the material firmly and, therefore, the inside of the cavity is either etched with an electric pencil or, when the die is to be used for welding large pieces of aluminium, grip marks are put in the cavity before the die is heat treated.

The gap between the two faces, or noses, of the die is also extremely important. If it is too large, the material will just collapse or bend away. This dimension is taken care of during manufacture and cannot be changed.

Finally, there is the offset of the die noses, which has the effect of making the weld look out of line around the circumference of the material. The purpose of the offset is to break the flash into two halves, so that removal is easy: otherwise the flash is likely to remain as a loose ring around the material and have to be cut off. The noses of the die also have to be sharp enough to virtually pinch off the flash around the weld, again to ensure that the complete flash can be easily removed.

The hardness and the temper of the die are most important as well. In the early days of cold welding, die breakage was very common and long after a machine was designed to weld 8mm copper rod, there were problems in containing the necessary forces within a die of this size.

PWM has been producing dies to extremely high standards and tolerances for over 30 years. As wire technology has improved, so has the demand for precision. PWM's on-going research and development programme has enabled it to produce dies that are capable of joining extremely fine wire. PWM was the first company outside the USA to develop a die that could be used in conventional cold welders to join wire as fine as 0.08mm (0.003145") in diameter. Individually hand-made in matched sets to the highest possible tolerances by skilled craftsmen, PWM's industry standard dies can now be produced for wire sizes between 0.08 (0.003145") and 6.50mm (0.256"). Dies can also be manufactured to suit round or profile wires and rods according to customers' specifications.

PWM dies can also be manufactured to suit various profiles, as long as the profile allows the die to be made in two halves - which is necessary for the removal of the welded wire - and the cross-sectional area is within the capacity of the machine.

It is also possible to weld two different wire sizes together. Generally, the larger diameter should not be more than 30% greater than the smaller. If the copper is considerably smaller in diameter than the aluminium, the copper will merely embed itself into the aluminium and no weld will be achieved.

PWM dies are of the industry standard type, hand
made in matched sets to the tightest possible tolerances.

Friction Welding

Friction welding (FRW) is a solid-state welding process that generates heat through mechanical friction between workpieces in relative motion to one another, with the addition of a lateral force called "upset" to plastically displace and fuse the materials. Because no melting occurs, friction welding is not a fusion welding process in the traditional sense, but more of a forge welding technique. Friction welding is used with metals and thermoplastics in a wide variety of aviation and automotive applications.

Benefits

The combination of fast joining times (on the order of a few seconds), and direct heat input at the weld interface, yields relatively small heat-affected zones. Friction welding techniques are generally melt-free, which mitigates grain growth in engineered materials, such as high-strength heat-treated steels. Another advantage is that the motion tends to "clean" the surface between the materials being welded, which means they can be joined with less preparation. During the welding process, depending on the method being used, small pieces of the plastic or metal will be forced out of the working mass (flash). It is believed that the flash carries away debris and dirt.

Another advantage of friction welding is that it allows dissimilar materials to be joined. This is particularly useful in aerospace, where it is used to join lightweight aluminium stock to high-strength steels. Normally the wide difference in melting points of the two materials would make it impossible to weld using traditional techniques, and would require some sort of mechanical connection. Friction welding provides a "full strength" bond with no additional weight. Other common uses for these sorts of bi-metal joins is in the nuclear industry, where copper-steel joints are common in the reactor cooling systems; and in the transport of cryogenic fluids, where friction welding has been used to join aluminium alloys to stainless steels and high-nickel-alloy materials for cryogenic-fluid piping and containment vessels.

Friction welding is also used with thermoplastics, which act in a fashion analogous to metals under heat and pressure. The heat and pressure used on these materials is much lower than metals, but the technique can be used to join metals to plastics with the metal interface being machined. For instance, the technique can be used to join eyeglass frames to the pins in their hinges. The lower energies and pressures used allows for a wider variety of techniques to be used.

Metal Techniques

Rotary Friction Welding

Rotary friction welding (RFW), for plastics also known as spin welding, uses machines that have two chucks for holding the materials to be welded, one of which is fixed and the other rotating.

In direct-drive friction welding (also called continuous drive friction welding) the drive motor and chuck are connected. The drive motor is continually driving the chuck during the heating stages. Usually, a clutch is used to disconnect the drive motor from the chuck, and a brake is then used to stop the chuck.

In inertia friction welding the drive motor is disengaged, and the work pieces are forced together by a friction welding force. The kinetic energy stored in the rotating flywheel is dissipated as heat at the weld interface as the flywheel speed decreases. Before welding, one of the work pieces is attached to the rotary chuck along with a flywheel of a given weight. The piece is then spun up to a high rate of rotation to store the required energy in the flywheel. Once spinning at the proper speed, the motor is removed and the pieces forced together under pressure. The force is kept on the pieces after the spinning stops to allow the weld to "set".

Linear Friction Welding

Linear friction welding (LFW) is similar to spin welding, except that the moving chuck oscillates laterally instead of spinning. The speeds are much lower in general, which requires the pieces to be kept under pressure at all times. This also requires the parts to have a high shear strength. Linear friction welding requires more complex machinery than spin welding, but has the advantage that parts of any shape can be joined, as opposed to parts with a circular meeting point. Another advantage is that in many instances quality of joint is better than that obtained using rotating technique.

In June 2016, the following materials could be welded: commercially pure copper (C101) /commercially pure aluminium (AA1050) /aerospace grade aluminium alloy (AA6082) /microalloyed steel (proprietary) /nickel alloy (Inconel 718) to conform a single part with all five materials joined as a demonstrator using LFW. Previously, a world-record weld interface area of 13,000 mm2 was successfully welded using similar materials welding: aluminium, steel and aerospace-grade titanium.

The most important parameters in the LFW process are Friction Pressure, Forging Pressure, Burn-off, Frequency, Amplitude, Stick out and perhaps their respective ramps or variation against time. The Friction Pressure is the one maintained between the parts to be welded during the oscillation period. The Forging pressure is the one kept for a short period of time after the oscillation is stopped and is typically around 20% over the Friction Pressure. The Burn-off is the linear measurement of the material "consumption" or transformed into "flash" (material that escapes around the welding). Frequency and Amplitude describe the movement of the oscillator and hence of one of the parts to be welded. Stick out is the linear measurement of the amount of material that the parts have protruding from the tooling (oscillator and forging tooling).

Friction surfacing

Friction surfacing is a process derived from friction welding where a coating material is applied to a substrate. A rod composed of the coating material (called a mechtrode) is rotated under pressure, generating a plasticised layer in the rod at the interface with the substrate. By moving a substrate across the face of the rotating rod a plasticised layer is deposited between 0.2–2.5 millimetres (0.0079–0.0984 in) thick depending on mechtrode diameter and coating material.

Thermoplastic Technique

Linear Vibration Welding

In linear vibration welding the materials are placed in contact and put under pressure. An external vibration force is then applied to slip the pieces relative to each other, perpendicular to the pressure being applied. The parts are vibrated through a relatively small displacement known as the amplitude, typically between 1.0 and 1.8 mm, for a frequency of vibration of 200 Hz (high frequency), or 2–4 mm at 100 Hz (low frequency), in the plane of the joint. This technique is widely used in the automotive industry, among others. A minor modification is angular friction welding, which vibrates the materials by torquing them through a small angle.

Orbital Friction Welding

Orbital friction welding is similar to spin welding, but uses a more complex machine to produce an orbital motion in which the moving part rotates in a small circle, much smaller than the size of the joint as a whole.

Seizure Resistance

Friction welding may unintentionally occur at sliding surfaces like bearings. This happens in particular if the lubricating oil film between sliding surfaces becomes thinner than the surface roughness, which may be due to low speed, low temperature, oil starvation, excessive clearance, low viscosity of the oil, high roughness of the surfaces, or a combination thereof.

The seizure resistance is the ability of a material to resist friction welding. It is a fundamental property of bearing surfaces and in general of sliding surfaces under load.

Explosive Welding

Joining of large-sized components of difficult-to-weld metals are welded by explosive welding. Strong metallurgical joints can be produced between parts of the same metal or dissimilar metals, for example, steels can be welded to tantalum though the melting point of tantalum is higher than the vaporisation point of steel.

In many of the critical components used in space and nuclear applications, explosive welding is used to fabricate them as they cannot be made by another process, and quite often this proves to be the least costly process in some of the commercial applications. However, most explosive welding is done on sections with relatively large surface areas though in some applications small components are also fabricated by this process.

Principle of Operation of Explosive Welding

The nature of the interface between the impacting components depends upon the velocity with which they strike against each other. A flat interface is formed if the collision velocity is below the critical value for a particular combination of materials being welded. Such welds are not considered good because small variation in the collision conditions can result in lack of bonding and thus an unacceptable weld.

Welds made with collision velocities above the critical value have a wavy interface with the amplitude of the waves varying between 0.1 and 4.0 mm and wavelength from 0.25 to 5.0 mm, depending upon the welding conditions. Welds with such an interface have better mechanical properties than those with flat interface.

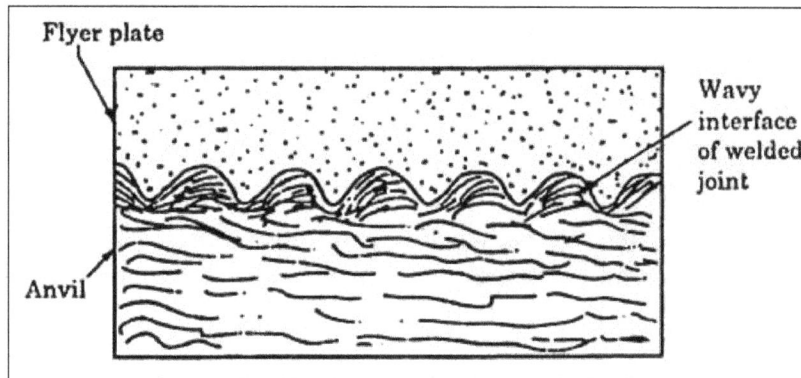

Explosive weld with wavy interface.

In such welds, a phenomenon known as surface jetting is also observed so that a small jet of metal is formed from the metals of the two impacting components. Such a jet is freely expelled at the edge of the joint, however, if it is trapped it results in rippling effect.

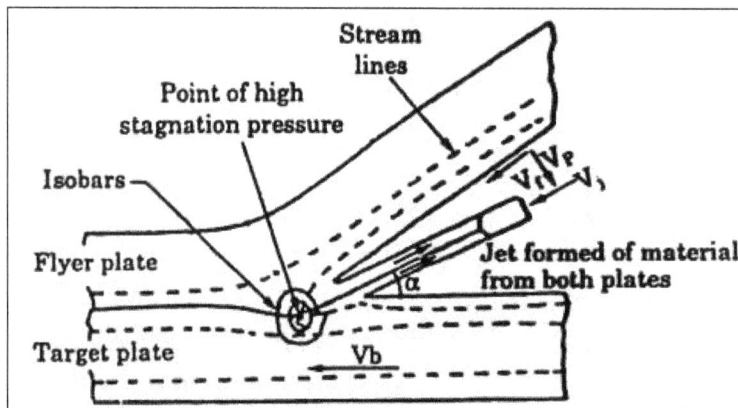

Jet formation in explosion welding.

In explosive welding setup the impact velocity becomes plate velocity Vp, and it must be high enough for the impact pressure to exceed the yield stress of the material by a considerable margin. The collision point velocity, Vcp i.e., the velocity at which the collision point moves along the surface being joined, must also be less than the velocity of sound in the two materials.

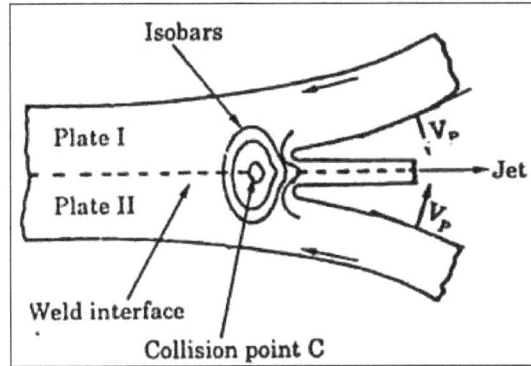

Jet formation in explosion welding under symmetrical
angular stand-off conditions.

The relationship between the different velocities is shown in the vector of wherein Vis the impact velocity, Vj, jet velocity, Vb the base plate velocity, and a is the angle of incidence which becomes actual stand-off angle g.

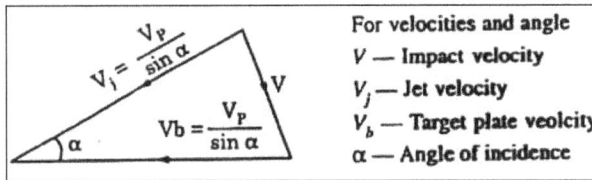

Velocity diagram for explosion welding.

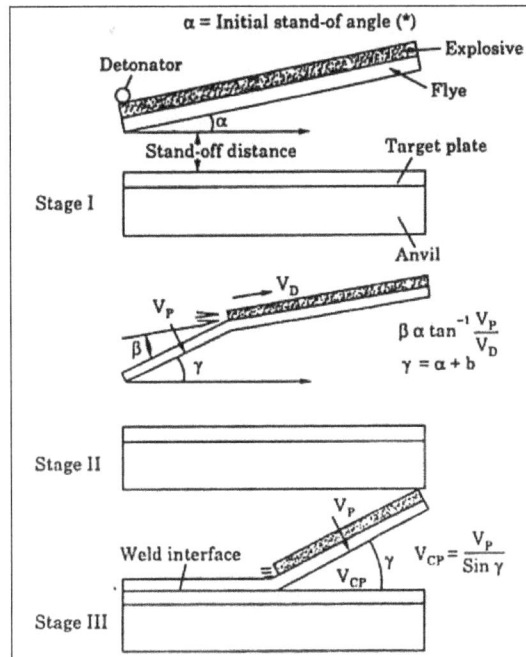

Different stages in explosion welding.

The explosive welds are made by either of the two setups. The welds are best made with parallel configuration of components in which only one plate is accelerated. In such a setup the detonation velocity of the explosive must be less than the velocity of sound in material to be joined in order to satisfy the condition that the collision point velocity, Vcp, must be subsonic. It is, however, difficult to fulfill this condition with most of the explosives as is evident from table.

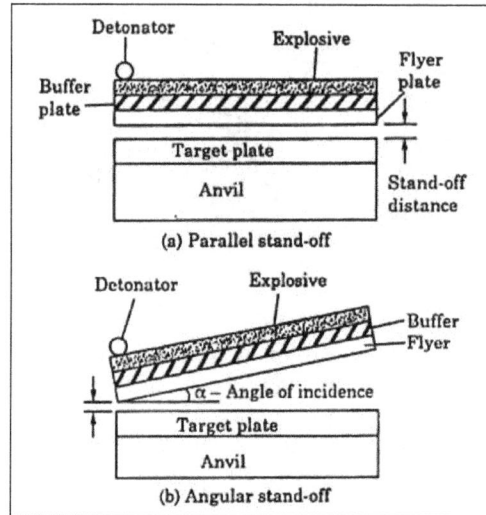

Component arrangement for parallel stand-off method and angular stand-off method.

Table properties of some well known explosives (After Tylecote)

Explosive	Density g/cm³	Detonation velocity m/ sec
TNT	1.56	6700
RDX	1.65	8200
PETN	1.70	8300
Tetryl	1.71	7900
Comp. B	1.68	7900
Detasheet (Du Pont)	1.40	7100
Metabel (I.C.I)	1.47	7000
ANFO*	0.85	

The detonation velocity of the explosive must be less than about 120% of the sonic velocity, Vs of the material being welded.

$$\text{where } V_s = \sqrt{\frac{k}{p}}, \text{ and } k = \frac{E}{1-2\sigma}$$

where, k = adiabatic bulk, dynes/cm²,

p = material density, gms/cm³

E = Young's modulus, and

σ = Poisson's ratio.

If the sonic velocity of the explosive is greater than 120% of the sonic velocity of the material with higher sonic velocity, a shock wave develops. This results in an extremely steep rise to the maximum pressure. (The maxim pressure at the interface is equal to the detonation pressure of the explosive).

In such a case, the material just in front of the shock wave experiences no pressure, while the material just behind the shock wave is compressed to peak pressure and density. The shock wave travels through the material at a supersonic velocity and creates significant plastic deformation locally and results in considerable hardening known as shock hardening.

The second type of detonation is when the detonation velocity is between about 100% and 120% of the sonic velocity of the material being welded. This results in a detached shock wave that travels slightly ahead of the detonation.

When the detonation velocity is less than the sonic velocity of the metal, the pressure generated by the expanding gases, and which is imparted to the metal, moves faster than the detonation. Though no shock wave is produced but the rising pressure reaches its peak value.

In cases 2 and 3, i.e., detached shock wave and no shock wave cases, pressure is generated ahead of the collision point of the metal plates. If a sufficiently large pressure is generated, it will cause the metal just ahead of the collision point to flow as a jet into the space between the plates. This high velocity jet effaces the material that removes the unwanted oxides and other unwanted surface films. At the collision point, the newly cleaned metal surfaces impact at high pressure, typically between 0.5 and 6 GPa.

Also, a significant amount of heat is generated upon detonation of the explosive. However, as detonation is completed within a few hundred microseconds so a very little part of it flows into the metal. Thus, no bulk diffusion takes place and a weld with only localised melting is produced.

It is therefore better to use angular setup in which the velocity of collision point is a function of the plate velocity and the initial stand off angle while it is only indirectly dependent on the detonation velocity VD, as is evident from the following relationship.

$$V_{cp} = \frac{V_p}{\sin\left[a + \tan^{-1}\left(\frac{V_p}{V_D}\right)\right]}$$

The plate velocity V_p is related to the mass of the plate and the explosive as well as the impulse (per unit mass) of the explosive. Knowing these parameters V_p can thus be calculated.

In the angular setup the wavelength of the ripples is directly related to the collision point velocity; while the shape of the ripples depends upon the plate velocity. Crested waves are most often produced with high plate velocity. For example, in welding aluminium with fixed stand-off angle, increasing the plate velocity from 260 m/sec to 410 m/sec results in a change from a sinusoidal wave formation to a highly tilted saw-tooth type of wave. Also, increasing the stand-off angle from 0.75° to 4.5° increased the wave length from 110 to 150 pm.

The pitch of the ripples also varies with the stand-off angle. No variation in waves was reported for welds in steel with angles between 1° and 15° but the pitch and amplitude increased with the angle. For a stand-off angle between 15° and 20° the interface became completely flat, above 20° no weld was produced.

The impact conditions for parallel plate setup are related by the following equation:

$$V_P = 2V_{cp} \sin\left(\frac{\gamma}{2}\right)$$

where V_{cp} is the impact or collision point velocity which is equal to the detonation velocity (V_D) of the explosive, y is referred to as the dynamic bend angle. It is the angle created between the flyer and target plates at the impact point, while V_p is the plate collision velocity at the point of impact.

Typically, the detonation velocity ranges between 1200 and 3800 m/sec depending upon the metal to be welded. The stand-off distance, which is an independent variable like V_D, is selected to achieve a specific dynamic bend angle and velocity of impact.

The dynamic bend angle is a dependent variable that is controlled by the detonation velocity (V_D) and the stand-off distance. Typical values for y are between 2 and 25 degrees. This results in a plate collision velocity at the impact point (V_p) of about 200 to 500 m/sec.

An important aspect of explosive welding is the flow pattern in the region of the collision point. Under conditions of subsonic flow the metal is reported to behave as a non-viscous compressible fluid. Due to jet formation oxide films and absorbed gases are completely removed from the weld. However, when the jet becomes unstable the gases and oxide films may get entrapped; this seems to occur with Reynold number in excess of 50. When the jet is entrapped it can either result in continuous molten metal layer of ½ – 250 pm thickness or in the formation of a rippled interface which often has localised fused zones on the forward side of the crest.

Methods of Operation of Explosive Welding

The target plate remains stationary and is often supported on an anvil of a large mass. When the explosive is detonated it thrusts the flyer plate towards the target plate. To protect the flyer plate from surface damage due to impaction as well as to control the collision point velocity, a thin layer of rubber or PVC or even chipboard is placed between it and the explosive to act as buffer or attenuator.

The explosive may be in sheet form but usually it is in granular form and is spread uniformly over the buffer plate. The force exerted by the flyer plate due to explosion depends upon the detonation characteristics and the quantity of the explosive. Welding is completed in microseconds with very little overall deformation, if any. Generally the welding operation is carried out in air but sometimes a rough vacuum of about 1 torr i.e., 1 mm of mercury or $133.322 \times 10^{-6} \, N/mm^2$ may be used.

For explosive welding it is required to impart subsonic velocity (V_p) to the flyer plate. This has to be done with an explosive which often has a fairly constant detonation velocity of about 6000 m/sec. The weight of the explosive required for a specific welding job is determined by trial and error, and there appears to be a linear relation between the ratio (weight of explosive/weight of flyer plate) and the flyer plate velocity, V_p. A ratio of 0.5 gives a plate velocity of 900 m/sec for Du Pont sheet explosive EL 506 D using a thin layer of rubber as buffer. For successful explosive welding it is required that the velocities of the two plates must be similar and this necessitates that the angle of inclination between them should be small. With low angles the impact velocity required to produce waves at the interface becomes greater.

When explosive welding is carried out at the normal atmospheric pressure, the gas between the plates provides the cushioning effect which not only necessitates higher minimum velocity but may also lead to inconsistent results. For welding aluminium in vacuum of about 1 mm of Hg the collision velocity should be about 150 to 300 m/sec with an included angle of 1 ° to 2°. In order to accelerate the plates being welded to this velocity the stand-off distance, should be equal to 1/4 to 1/2 times the plate thickness as marked.

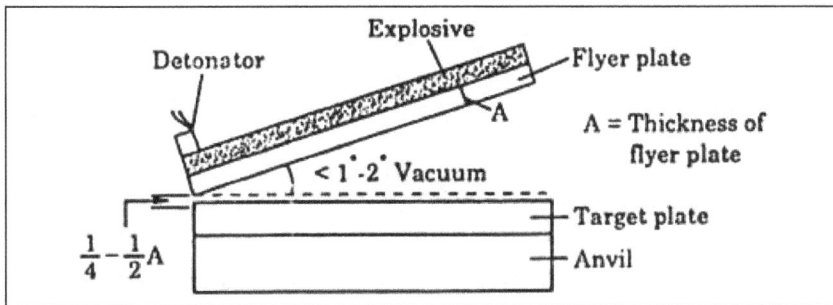

Angular stand-off method with small angle of inclination to
keep the velocities of two plates nearly equal during explosion welding.

The stand-off distance is held by the use of a shim. There are many types of shims which are designed to be consumed by the jet so as not to adversely affect the weld.

If the effective angle attained by the flyer plate is too small, the velocity will be highly supersonic and no waves will be formed at the interface. Ideally the detonation velocity of the explosion should be subsonic. However, it is rarely possible in practice as detonation velocities exceed 5500 m/ sec while the velocity of sound in steel which is among the highest among metals, is only 5200 m/sec.

Table. Velocities of sound in metals

Material	Velocity of sound (m/sec)
Lead	2100
Copper	4200
Steels	4800-5200
Aluminium	5500

No special surface cleaning treatment is required for explosive welding; however grease, if present, in the surface must be removed. Dirt or oxide if present in excess will get accumulated near the crests of the ripples and may lead to reduced strength of the joint.

The pressure corresponding to a plate velocity of 120 m/sec on copper is 2400 N/mm² and for a velocity of 220 m/sec on aluminium it is 6200 N/mm². These pressures are adequate to force metal through cracks in the oxide film and to weld it. It is also reported that even when the surfaces of 18/8 stainless steel and mild steel were covered with adherent layer of black oxide they were welded satisfactorily with the desired rippled interface.

Process Variables in Explosive Welding

The major process variables in explosive welding are:

- Impact velocity,
- Stand-off distance,
- Angle of approach.

Impact Velocity

The impact velocity depends upon the ratio of the weight of the explosive to that of the weight of the flyer plate and also on the contact angle. For each material there is a minimum velocity below which welding does not take place, for example, copper cannot be welded with velocities below 120 m/sec and aluminium at velocities less than 255 m/sec.

The maximum velocity that can be usefully employed for explosive welding is decided by the velocity of sound in the target plate material because at supersonic velocities the wave in the target cannot propagate ahead of the bonding front. Also, the velocity near the edge of the workpiece is reduced resulting in relieving of pressure in such zones; this may lead to unsatisfactory welding near the work edges when near-minimum velocity is employed.

The minimum velocity for any material is determined by the magnitude at which the projectile material becomes sufficiently plastic on impact to form a divided jet. Different explosives result in different velocities and thus due consideration need be given while selecting the explosive.

Two important properties of explosives for welding are, detonation velocity and hazard sensitivity. The latter affects the handling safety as it refers to the thermal stability, storage life, and shock sensitivity of the explosive.

Whereas detonation velocity is proportional to the density of the explosive, the pressure generated is proportional to both the density and the detonation velocity. The detonation velocity of an explosive depends upon its thickness, packing density as well as the passive material mixed with the explosive to decrease its detonation velocity.

Some of the explosives popularly employed for giving the desired detonation rates include:

- Ammonium nitrate-TNT-atomised aluminium mixture,
- Ammonium nitrate pallets with 6 to 12% diesel fuel,
- Nitroguanidinne plus inert material,
- Amatol and sodatol with 30 to 55% rock salt.

Stand-off Distance

Increasing the stand-off distance increases the angle of approach between the flyer plate and the target plate. This results in increased size of the wave which reaches a maximum and then decreases as the stand-off distance is further increased. In a parallel set-up a stand-off distance of between ½ and 2 times the thickness of the flyer plate is normally used; the lesser stand-off distance is used with an explosive having high detonation velocity.

Angle of Approach

For successful explosive welding the angle of impact or approach is usually required to be between 5° and 25°. With a parallel setup this angle can develop only if there is a proper stand-off distance. When welding tube-to-tubeplate, a suitable angle is achieved by tapering the hole in the tube plate.

Weld Joint Properties of Explosive Welding

Joint properties of an explosive weld are affected depending upon whether the interface is formed by trapped jet which results in rippling, or the free jet that results in the total expulsion of a thin interfacial layer. The trapped jet technique is preferred as it results in extended interface to the extent of almost 75 % in length.

It is reported that fused nuggets are found imbedded in front and in some instances just behind the crest of the interfacial wave formation. In these zones there appears to be considerable mixing of dissimilar metals leading to detached particles of one metal in the other, or to the production of solid solutions or intermetallic compounds. Free jetting may give a continuous cast interfacial zone such as in copper. Free jetting is capable of causing complete expulsion of the interfacial metallic zone.

On aluminium a 10° stand-off angle may result in almost invisible solid-state interface, all traces of which can be removed by annealing, while a parallel stand-off gives a rippled interface with a dark interfacial layer which remains unaffected by annealing.

Interfacial hardness of welds in copper increased from 65 to 150 VHN while mild steel to copper welds resulted in more hardening in the copper then the steel while copper hardened from 60 to 160 VHN, the steel hardened from 120 to 160 VHN. Stainless steel reached a hardness value of 400 VHN possibly due to the formation of martensite while copper to which it was welded increased in hardness from 60 to 150 VHN.

It is evident that non-equilibrium phases can be produced during explosive welding and that high strain rates result in very high diffusion rates; also that the phases produced are sensitive to the exact method of operation and the process variables used.

Variants of Explosive Welding

Explosive spot welding is perhaps the only variant of the process. In this process a small explosive charge is used to join difficult-to-weld metals,

A sturdy and compact hand-held explosive spot welder weighing about 5 kg can be employed to produce welds up to about 10 mm in diameter. Electric current is employed for igniting the charge and the unit is provided with multiple safety interlocks. PTN (pentaery thritetranitrate) explosive capsules of different weights are available for use with the standard cap.

Usually explosive is in direct contact with the workpiece to be welded. However, plastic buffer discs can be provided to protect the work surface, where necessary. Stand-off distance can be varied if required, but the normal practice is to control the explosive force by using as small an explosive charge as possible.

Most of the engineering metals can be spot welded by explosive welding but the process has been reported to be particularly successful for welding austenitic stainless steel to cobalt-base alloys for use in high temperature applications and also for joining nickel-base alloy such as Inconel and nickel. Aluminium alloys can also be spot welded easily provided they are cleaned of the tenacious oxide layer a maximum of 4 hours prior to welding.

Explosive spot welding may prove indispensable for space applications such as emergency repairs to spacecraft or even for the erection of devices in space.

Applications of Explosive Welding:

Explosive welding is a specialised process used for lap joints in difficult to weld metals and their combinations. Aluminium and copper can be welded to stainless steel, aluminium to nickel alloys, and stainless steel to nickel. Aluminium can be welded to copper and stainless steel to brass. The bonding of aluminium to steel is complicated by the formation of $FeAl_2$ layer at the interface.

However, this can be remedied by interposing an intermediate layer of a metal compatible to both these metals, or by selecting the parameters so as to reduce the extent of diffusion that occurs across the interface. The strength of welds depends on the structure at the interface but a weld that does not have a brittle interface usually gives 100 percent efficiency in – shear or tension.

In general metals with elongation of at least 5% in 50 mm gauge length and charpy V-notch impact strength of 13.5 joules or more can be welded by explosive welding. Normally strength and hardness increase and ductility decreases as a result of explosive welding. This is caused by severe plastic deformation encountered particularly in the flyer plate. Explosive welding may also increases the ductile-to-brittle transition temperature of carbon steel.

Cladding of plates is one of the major commercial applications of explosive welding. Clad plates are supplied in as welded condition because the increased interfacial hardness does not effect the engineering properties of plates. Slight distortion of plates may take place during cladding which need be rectified to meet the standard flatness specifications. Rollers or a press may be employed for the purpose.

Cladding of cylinders both inside and outside is done by explosive welding; one application of this is the internal cladding of steel forgings with stainless steel to make nozzles, 12 mm to 600 mm diameter and up to 900 mm length, for connection to heavy-walled pressure vessels.

Metals which are incompatible for fusion welding are welded by using transition welds made by explosive welding.

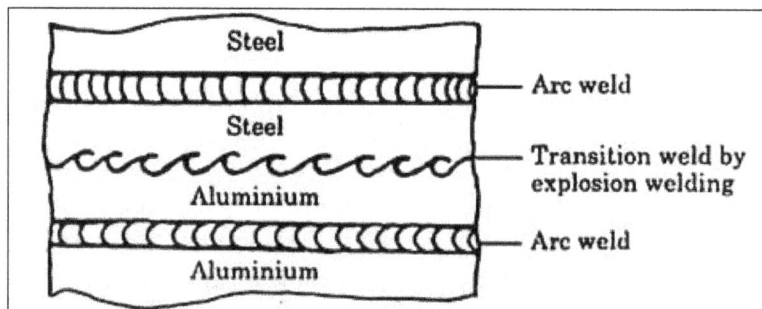

Welding of incompatible material using transition joint.

Transition joints cut from thick explosive welded plate of aluminium and steel or aluminium and copper provide efficient conductors of electricity. This technique is also used for fabrication of anodes for primary aluminium steel in tubes ranging in diameter from 50 to 300 mm. Other metals joined by this technique include titanium to steel, zirconium to stainless steel, zirconium to nickel base alloys, and copper to aluminium.

Explosive welding also finds an application in the fabrication of heat exchangers where tube-to-tubeplate joints can be made by this process. A small explosive charge is used to make the joint as sown in three steps in the figure. Tubes may be welded individually or in groups, the number of tubes welded at a time depends upon the quantity of explosive that can be exploded safely in single detonation.

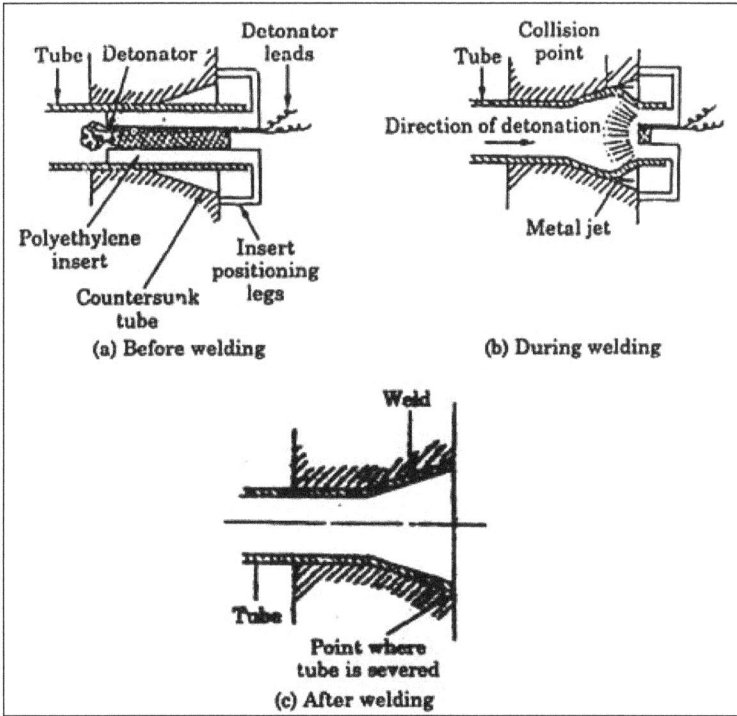

Different stages in explosion welding of tube-to-tubeplate.

The schematic of the overall set-up for explosion welding of plugs for sealing the leaking tubes, through remote control.

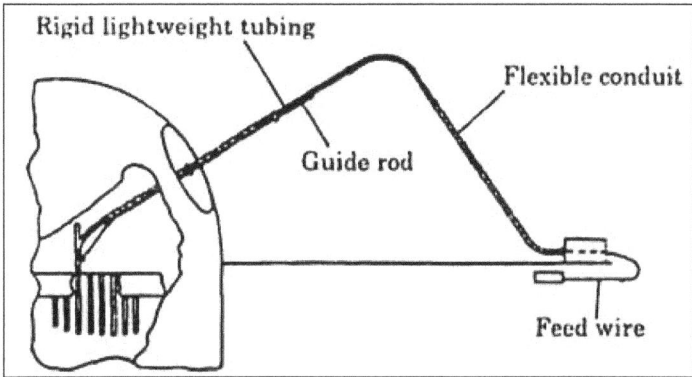

Schematic representation of explosion welding of
plugs for sealing the tubes through remote control.

Tubes welded in tube-to-tubeplate joints are usually of diameter between 12 to 40 mm. Metals welded for such joints include, steel, copper alloys, stainless steels, nickel alloys, clad steels, and both aluminium and titanium to steels.

Explosive welding can be used for repair and building-up particularly both inside and outside of cylindrical components.

Types

This welding can be classified into two types according to the setup configuration.

Oblique Explosion Welding:

In this type of welding process base plate is fixed on an anvil and filler plate makes an angle with the base plate. This welding configuration is used to join thin and small plates.

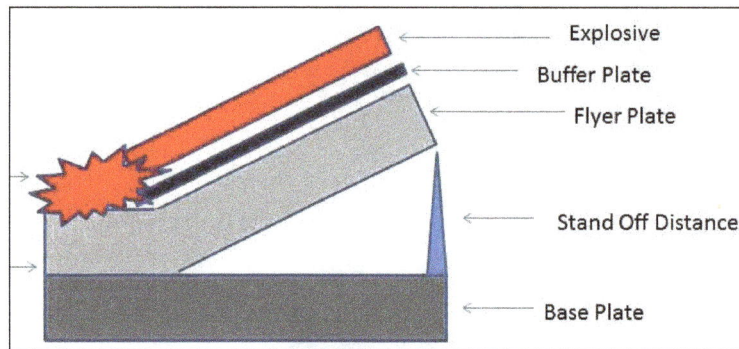

Explosion Welding.

Parallel Explosion Welding

As the name implies, in this welding configuration filler plate is parallel to the base plate. There is some standoff distance between base plate and flyer plate. This configuration is used to weld thick and large plates.

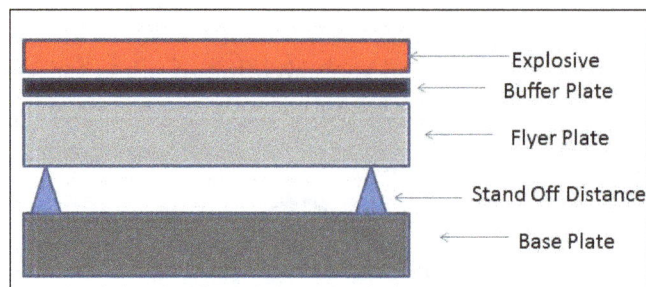

Parallel Explosion Welding.

Advantages and Disadvantages

Advantages

- It can join both similar and dissimilar material.

- Simple in operation and handling.

- Large surface can be weld in single pass.

- High metal joining rate.

- Mostly time is used in preparation of the welding.

- It does not effect on properties of welding material.

- It is solid state process so does not involve any filler material, flux etc.

Disadvantages

- It can weld only ductile metal with high toughness.

- It creates a large noise which produces noise Pollution.

- Welding is highly depends on process parameters.

- Higher safety precautions involved due to explosive.

- Designs of joints are limited.

Diffusion Bonding

Diffusion bonding is a joining process where in the principal mechanism for joint formation is solid state diffusion. The solid phase diffusion bonding of the product is used in various industrial fields to make making to high performance and making to a high function near-net shape in addition. Diffusion bonding offers many advantages, mainly the strength of the bonding line, which is equal to the base metals. The microstructure at the bonded region is exactly the same as the parent metals. On the other hand, this advantage joining process requires several strictly controlled condition: clean and smooth contacting surfaces which are free from oxides, etc., high temperature condition to promote diffusion process. In diffusion bonding, the bond strength is achieved by pressure, temperature, time of contact, and cleanness of the surfaces. The strength of the bond is primarily due to diffusion rather than any plastic deformation. Diffusion bonding is an attractive manufacturing option for joining dissimilar metals and for making the component with critical property continuity requirements. Unlike other joining processes the diffusion bonding process preserves the base metal microstructure at the interface. More importantly no localized thermal gradient is present to induce distortion or to create residual stresses in the component.

This property results in a union where the joint is metallurgically and detectable, i.e., grain boundaries are not confined to the original joint face. For practical purposes the intimate contact and atomic exchange is assisted by heat and pressure from an external source, although no melting of the material takes place. Bond strengths up to parent material properties are achievable. The joining aspect of the process is similarly concerned with elevated temperature flow properties and fine grain sizes. In achieving intimate contact of two originally free surfaces, diffusion accounts for only a small, though vital amount of the mass transport required, the majority being achieved by plastic deformation. Thus the low flow stresses associated with fine grain sizes are desirable for bonding just as for superplastic forming. Again, in-process grain growth can adversely affect the diffusion bonding process and must be of special concern for bonds to be achieved in the process cycle.

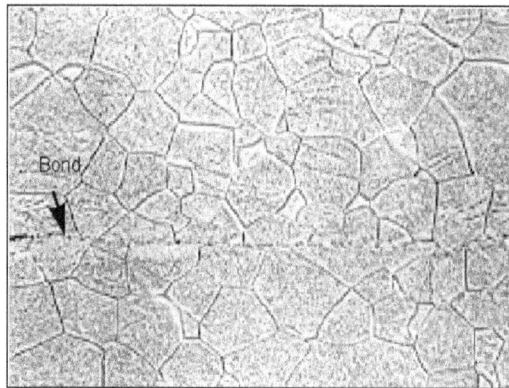

Micro Structure of Diffusion Bonding

Some metals will unite to form a homogeneous structure when placed in intimate contact under temperature and pressure:

- Similar metals may be joined directly to form a solid-state weld. In this situation-required pressures, temperatures, times are dependent only account the characteristics of the metals to be joined and their surface preparation.

- Similar metals can be joined with a thin layer of a different metal between them. In this case, the layer may promote more rapid diffusion or permit increased micro deformation at the joint to provide more complete contact between the surfaces. This interface metal may be diffused into the base metal by suitable heat treatment until it no longer remains a separate layer.

- Two dissimilar metals may be joined directly where diffusion-controlled phenomena occur to form a bond.

- Dissimilar metals may be joined with a third metal between the faying surfaces to enhance weld formation either by accelerating diffusion or permitting more complete initial contact in a manner similar to category.

Two necessary conditions that must be met before a satisfactory diffusion weld can be made are:

- Mechanical intimacy of metal-to-metal contact must be achieved.

- Interfering surface contaminants must be disrupted and dispersed to permit metallic bonding to occur.

One property of a correctly prepared surface is its combined flatness and smoothness. A certain minimum degree of flatness and smoothness is required to assure uniform contact. A secondary affect of machining or abrading is the cold work introduced into the surface. Recrystallization of the cold worked surfaces tends to increase the diffusion rate in the weld region across the interface between them. The need for oxide removal is apparent because it prevents metal-tometal contact.

A three stages mechanistic model, adequately describes weld formation:

- In the first stage, deformation of the contacting asperities occurs primarily by yielding and by creep deformation mechanism to produce intimate contact over a large fraction of the interfacial is. At the end this stage, the joint is essentially a grain boundary at the areas of contact with voids between these areas.

- During the second stage, diffusion become more important than deformation, and many of the voids disappear as grain boundary diffusion of atoms continues.

- In the third stage, the remaining voids are eliminated by volume diffusion of atoms to the void surface.

Two types of diffusion bonding processing have been investigated in the development of low cost structure. These are:

- Massive diffusion bonding.

- Thin sheet diffusion bonding Massive diffusion bonding is a process used in the manufacture of heavy structure from plate elements.

This is essentially a mechanical bonding process as illustrated in. The essential advantages of this manufacturing technique are its ability to produce heavy sections with a much-improved material utilization relative to conventional processes such as machining from solid.

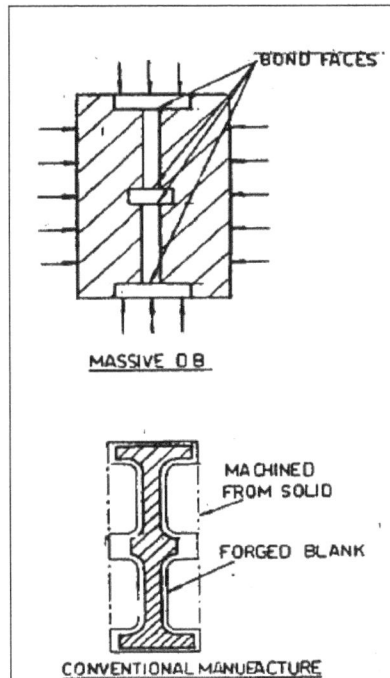

Massive Diffusion Bonding.

The most common diffusion bonding process practiced in airframe structure manufacture is thin sheet diffusion bonding because of the large area bonds associated with thin sheet structures and the fact that the sheets are at Superplastic Forming (SPF) temperature, bonding is affected by means of inert gas pressure applied in a bonding tool as illustrated in. This process ensures uniform pressure over the whole bond area and enables mismatch between the mating faces to be overcome by the SPF properties of the material.

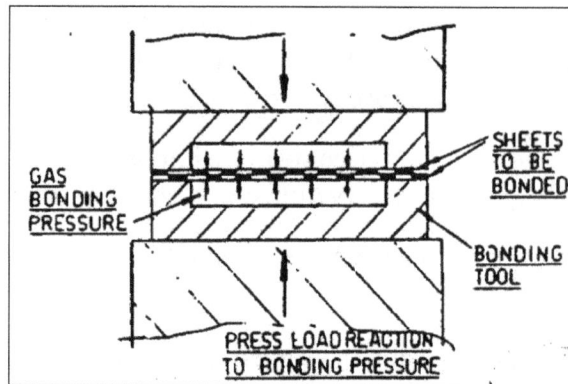

Thin Sheet Diffusion Bonding.

The variables of diffusion bonding are:

- Temperature is the most influential variable since it determine the extent of contact area during stage one and the rate of diffusion which governs void elimination during the second and third stages of welding.

- Pressure is necessary only during the first stage of welding to produce a large area of contact at the joining temperature. Removal of pressure after this stage does not significantly affect joint formation. However, premature removal of pressure before completion of the first stage is detrimental to the process.

- Rough initial surface finishes generally adversely affect welding by impeding the first stage and leaving large voids that must be eliminated during the later stages of welding.

- The time required to form a joint depends upon the temperature and pressure used; it is not an independent variable.

Metallurgical Factors

Two factors of particular importance with similar metal weld are allotropic transformation and micro structural factors that tend to modify diffusion rates. Allotropic transformation (phase transformation) occurs in some metals and alloys. The important of the transformation is that the metal is very plastic during that time. This tends to permit rapid interface deformation at lower pressures in much the same manner as does recrystallization. Diffusion rates are generally higher in plastically deformed metals as they recrystalize. Another means of enhancing diffusion is alloying or more specially, introducing elements with high diffusivity into the systems at the interface. The function of a high diffusivity element is to accelerate void elimination. Alloying must be controlled to avoid melting at the joint interface. When using a diffusionactivated system, it is desirable to heat the assembly for some minimum time either during or after the welding process to disperse the high diffusivity element away from the interface. If this is not done, the high concentration of the element at the joint may produce metallurgically unstable structures.

Diffusion bonding is often combine with superplastic forming (SPF) for aerospace titanium structures. Combining the two processes of SPF and diffusion bonding in particular as concurrent processes, provides a potential for considerable cost and weight saving when compared with conventional fabricated structures which are typical of aerospace structures.

Cost saving accrue from the ability to form complex structure from simple starting blanks (in most cases flat blanks) and to form this into a complete structure in one operation. This means of manufacture significantly reduce the parts count relative to fabricate structures.

Some of the advantages of diffusion bonding process are:

- Joint can be produced with properties and microstructures very similar to those of the base metal. This is particularly important for light weight fabrications.

- Component can be joined with minimum distortion and without subsequent machining or forming.

- Dissimilar alloys can be joined that are not weldable by fusion processes or by processes requiring axial symmetry.

- A large number of joints in an assembly can be made simultaneously.

- Components with limited access to be joints can be assembled by these processes.

- Large components of metals that required extensive preheat for fusion welding can be joined by theses processes.

- Defects normally associated with fusion welding are not encountered.

- Economic advantages:

   ○ Simple starting blank form (particularly significant for titanium).

   ○ High material utilization.

   ○ Reduces parts count.

   ○ Process times which are insensitive to size, complexity of structural form, or number of components manufactured in one operation.

- Weight advantages:

   ○ These weight saving occur from the ability of SPF/DB in particular, to produce efficient structural forms with the elimination of fasteners and associated joint flanges.

The limitations of diffusion bonding process are:

- Generally, the duration of the thermal cycle is longer than that of conventional welding and brazing processes.

- Equipment costs are usually high and this can limit the size of components that can be produced economically.

- The processes are not adaptable to high production applications, although a number of assemblies may be processed simultaneously.

- Adequate non destructive inspection techniques for quality assurance are not available, particularly those that assure design properties in the joint.

- Suitable filler metals and procedures have not been yet developed for all structural alloys.

- The surfaces to be joined and the fit-up of mating parts generally required greater care in preparation than for conventional hot pressure welding or brazing process.

- The need to simultaneously apply heat and a high compressive force in the restrictive environment of a vacuum or protective atmosphere is a major equipment problem with diffusion welding.

Platelet Diffusion Bonding and its Application

Platelet diffusion bonding process involves precise photo etching or lasers cutting of thin platelets to the designed channel configuration.

Subsequently, the etched platelets are arranged and stacked together and diffusion bonded at elevated temperature. The diffusion of the element occurs at the platelet interface and results in a metallurgical bond joint. The bonded platelet panels are then formed and/or machined to the final hard ware configuration. Platelet diffusion bonding has been successfully applied to the wide range of engineering materials, such as stainless steels, copper, aluminium and titanium alloys, and refractory materials. This process offers a significant cost reduction in the production of fluid or gas flow devices with extremely small flow channels, particularly for aerospace and electronic application. The fabrication of a copper liner for a liquid rocket combustion chamber, which requires extensive cooling during operation in a severe hot gas environment. Flat panels were fabricated by diffusion bonding of thin copper platelets with etched channels, and then formed to chamber configuration.

Similar process was used to fabricate a stainless steel window frame in the fore body of a landbased missile. Cooling of the sapphire window is required to protect the electronic sensor underneath due to severe hypersonic flight environments. The temperature of the sapphire window must be uniformly controlled, because any temperature gradients in the window can cause a shift in the apparent target location and can blur or distort the target signal. Platelet diffusion bonding technology offers unique design and fabrication process producing extremely small and complicated cooling channels assuring a uniform temperature as required.

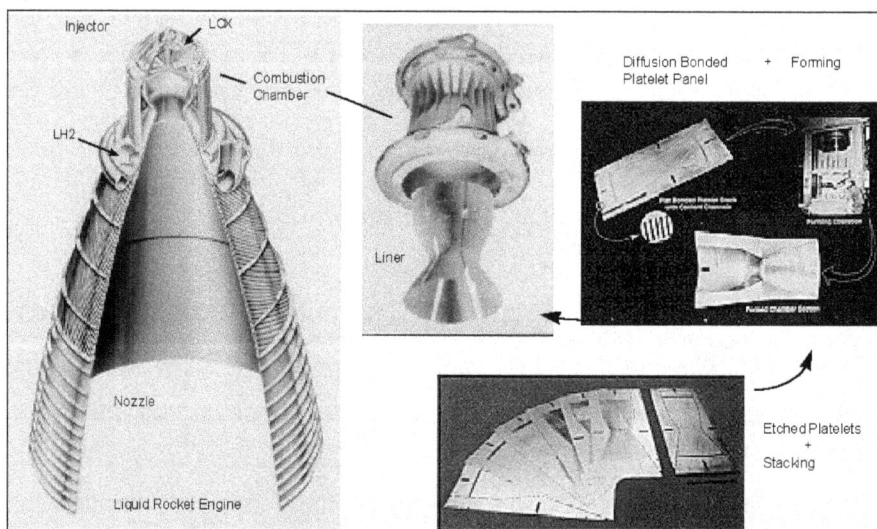

Fabrication of platelet liner of a liquid rocket combustion chamber.

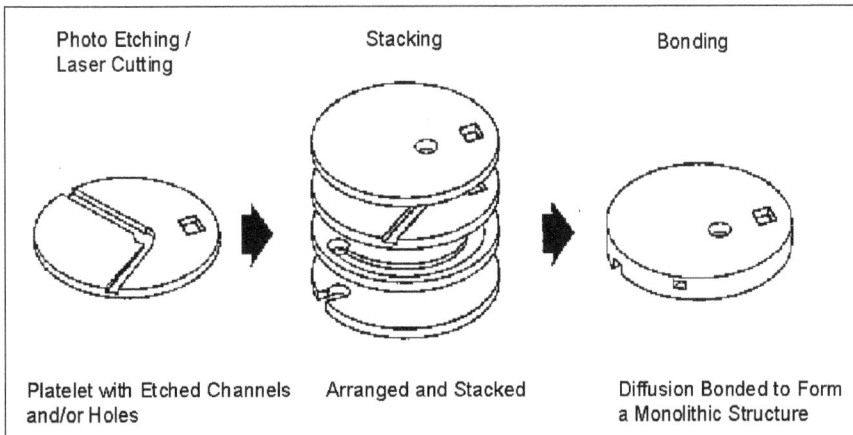

Platelet diffusion bonding concept.

In the assembly process, hot isostatic diffusion bonding or brazing could be used to joint the platelet panels to each other and to the structural support jacket of a high strength cast alloy. Internal channels in the liner are designed to allow the flow of liquid hydrogen for cooling the combustion chamber.

Advantages and Limitations

Advantages

- This solid state process avoids pitfalls of fusion welding.
- Dissimilar materials welds are possible.
- Properties and microstructures remain similar to those of base metals.
- Multiple welds can be made in one setup at the same time.
- Produces a product finished to size and causes minimal deformation.
- Presents less shrinkage and stresses compared to other welding processes.
- Highly automated process does not need skillful workforce.

Limitations

- Costly equipment especially for large weldments.
- Costly preparation with smooth surface finish and exceptional cleanliness.
- Protective atmosphere or vacuum required.
- Long time to completion.
- Not suited to high production rates.
- Difference in thermal expansion of members may need special attention.
- Limited nondestructive inspection methods available.

References

- Top-5-solid-state-welding-processes-metallurgy, welding: yourarticlelibrary.com, Retrieved 19 March, 2019

- What-is-cold-pressure-welding, home: coldpressurewelding.com, Retrieved 20 April, 2019

- Skowrońska, Beata; Siwek, Piotr; Chmielewski, Tomasz; Golański, Dariusz (2018-05-10). "Zgrzewanie tarciowe ultradrobnoziarnistej stali 316L". Przegląd Spawalnictwa - Welding Technology Review. 90 (5). Doi:10.26628/ps.v90i5.917. ISSN 2449-7959

- Explosive-welding-applications-and-variants-metallurgy, explosive-welding, welding, yourarticlelibrary.com, Retrieved 21 May, 2019

- Diffusion-welding: welding-advisers.com, Retrieved 23 July, 2019

5

Resistance Welding

Resistance welding uses the electrical resistance of materials which results in heat to form the weld. There are various types of resistance welding such as spot welding, flash welding, seam welding, projection welding, etc. This chapter has been carefully written to provide an easy understanding of resistance welding and its types.

Resistance welding or electric resistance welding is a welding technology widely used in manufacturing industry for joining metal sheets and components. The weld is made by conducting a strong current through the metal combination to heat up and finally melt the metals at localized point(s) predetermined by the design of the electrodes and/or the workpieces to be welded. A force is always applied before, during and after the application of current to confine the contact area at the weld interfaces and, in some applications, to forge the workpieces.

Parameters in Resistance Welding

The principle of resistance welding is the Joule heating law where the heat Q is generated depending on three basic factors as expressed in the following formula:

$$Q = I^2 Rt$$

where I is the current passing through the metal combination, R is the resistance of the base metals and the contact interfaces, and t is the duration/time of the current flow.

The principle seems simple. However, when it runs in an actual welding process, there are numerous parameters, some researchers had identified more than 100, to influence the results of a resistance welding. In order to have a systematic understanding of the resistance welding technology, we have carried out a lot of experimental tests and summarized the most influential parameters into the following eight types:

Welding current

The welding current is the most important parameter in resistance welding which determines the heat generation by a power of square as shown in the formula. The size of the weld nugget increases rapidly with increasing welding current, but too high current will result in expulsions and electrode deteriorations. The figure below shows the typical types of the welding current applied in resistance welding including the single phase alternating current (AC) that is still the most used in production, the three phase direct current (DC), the condensator discharge (CD), and the newly developed middle frequency inverter DC. Usually the root mean square (RMS) values of the welding current are used in the machine parameter settings and the process controls. It is often the tedious job of the welding engineers to find the optimized welding current and time for each individual welding application.

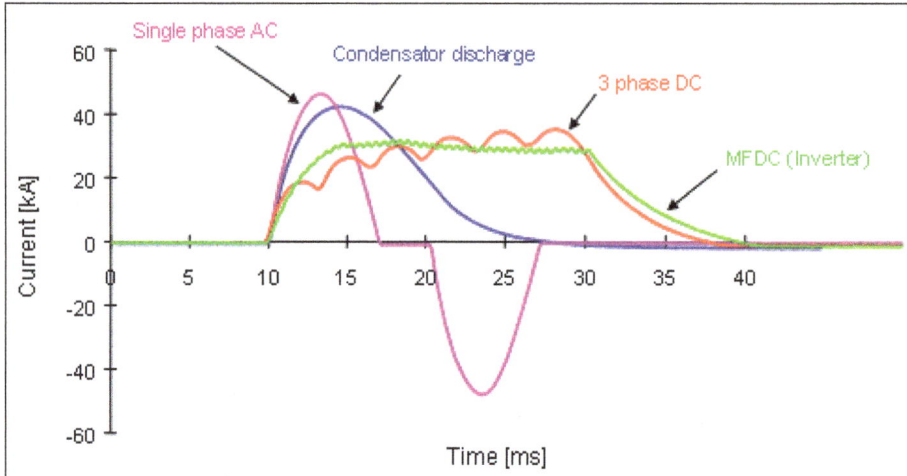

Welding Time

The heat generation is directly proportional to the welding time. Due to the heat transfer from the weld zone to the base metals and to the electrodes, as well as the heat loss from the free surfaces to the surroundings, a minimum welding current as well as a minimum welding time will be needed to make a weld. If the welding current is too low, simply increasing the welding time alone will not produce a weld. When the welding current is high enough, the size of the weld nugget increases with increasing welding time until it reaches a size similar to the electrode tip contact area. If the welding time is prolonged, expulsion will occur or in the worst cases the electrode may stick to the workpiece.

Welding Force

The welding force influences the resistance welding process by its effect on the contact resistance at the interfaces and on the contact area due to deformation of materials. The workpieces must be compressed with a certain force at the weld zone to enable the passage of the current. If the welding force is too low, expulsion may occur immediately after starting the welding current due to fact that the contact resistance is too high, resulting in rapid heat generation. If the welding force is high, the contact area will be large resulting in low current density and low contact resistance that will reduce heat generation and the size of weld nugget. In projection welding, the welding force causes the collapse of the projection in the workpiece, which changes the contact area and thereby the contact resistance and the current density. It further influences the heat development and the welding results.

Contact Resistance

The contact resistance at the weld interface is the most influential parameter related to materials. It however has highly dynamic interaction with the process parameters. The figure below shows the measured contact resistance of mild steel at different temperatures and different pressures. It is noticed that the contact resistance generally decreases with increasing temperature but has a local ridge around 300 °C, and it decreases almost proportionally with increasing pressure. All metals have rough surfaces in micro scale. When the welding force increases, the contact pressure increases thereby the real contact area at the interface increases due to deformation of the rough surface asperities. Therefore the contact resistance at the interface decreases which reduces the heat generation and the size of weld nugget. On the metal surfaces, there are also oxides, water vapour, oil, dirt and other contaminants. When the temperature increases, some of the surface contaminants (mainly water and oil based ones) will be burned off in the first couple of cycles, and the metals will also be softened at high temperatures. Thus the contact resistance generally decreases with increasing temperature. Even though the contact resistance has most significant influence only in the first couple of cycles, it has a decisive influence on the heat distribution due to the initial heat generation and distribution.

Materials Properties

Nearly all material properties change with temperature which add to the dynamics of the resistance welding process. The resistivity of material influences the heat generation. The thermal conductivity and the heat capacity influence the heat transfer. In metals such as silver and copper with low resistivity and high thermal conductivity, little heat is generated even with high welding current and also quickly transferred away. They are rather difficult to weld with resistance welding. On the other hand, they can be good materials for electrodes. When dissimilar metals are welded, more heat will be generated in the metal with higher resistivity. This should be considered when designing the weld parts in projection welding and selecting the forms of the electrodes in spot welding. Hardness of material also influences the contact resistance. Harder metals (with higher yield stress) will result in higher contact resistance at the same welding force due to the rough surface asperities being more difficult to deform, resulting in a smaller real contact area. Electrode materials have also been used to influence the heat balance in resistance welding, especially for joining light and non-ferrous metals.

Surface Coatings

Most surface coatings are applied for protection of corrosion or as a substrate for further surface treatment. These surface coatings often complicate the welding process. Special process parameter adjustments have to be made according to individual types of the surface coatings. Some surface coatings are introduced for facilitating the welding of difficult material combinations. These surface coatings are strategically selected to bring the heat balance to the weld interface. Most of the surface coatings will be squeezed out during welding, some will remain at the weld interface as a braze metal.

Geometry and Dimensions

The geometry and dimensions of the electrodes and workpieces are very important, since they influence the current density distribution and thus the results of resistance welding. The geometry of electrodes in spot welding controls the current density and the resulting size of the weld nugget. Different thicknesses of metal sheets need different welding currents and other process parameter settings. The design of the local projection geometry of the workpieces is critical in projection welding, which should be considered together with the material properties especially when joining dissimilar metals. In principle, the embossment or projection should be placed on the material with the lower resistivity in order to get a better heat balance at the weld interface.

Welding Machine Characteristics

The electrical and mechanical characteristics of the welding machine have a significant influence on resistance welding processes. The electrical characteristics include the dynamic reaction time of welding current and the magnetic/inductive losses due to the size of the welding window and the amount of magnetic materials in the throat. The up-slope time of a welding machine can be very critical in micro resistance welding as the total welding time is often extremely short. The magnetic loss in spot welding is one of the important factors to consider in process controls. The mechanical characteristics include the speed and acceleration of the electrode follow-up as well as the stiffness of the loading frame/arms. If the follow-up of the electrode is too slow, expulsion may easily occur in projection welding. The figure below shows measured process parameters in a projection welding process, which include the dynamic curves of the welding current, the welding force and the displacement of the electrode, where the sharp movement corresponds to the collapse of the projection in the workpiece.

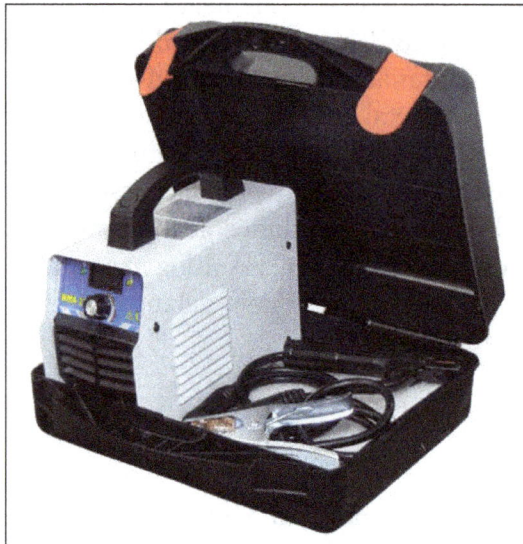

Electrode Degradation and Tip Dressing

The resistance welding process is characterized with a high current passing through the materials to be welded between the electrodes under pressure for generating concentrated heat to form a weld. This highly concentrated heat also causes problems to the electrode tips with increasing number of welds.

Mechanisms of Electrode Degradation

The severe conditions of high current and pressure during resistance welding expose the electrode tips at a high risk of degradation. The photo to the right shows a comparison of the new and used electrode tips in spot welding of galvanized steel sheets. With increasing number of welds, there will be two major changes in the electrode tips:

- Geometric changes: The electrode tip diameter will increase due to deformation and wear, such as mushrooming, pitting or local material removal by picking up.

- Metallurgical changes: The material properties near the tip surface will also change during resistance welding such as alloying with sheet and coating materials, and recrystallization and softening by overheating.

Effects of Electrode Degradation

The increasing tip diameter will result in larger contact area between electrode and sheet thereby reducing the current density passing through the weld interface. At the same time, alloying of the electrode material with sheet and coating materials at the tip surface will reduce the conductivity of the electrode tip thereby also drag heat concentration away from the weld interface. Both effects lead to progressively reducing weld nugget sizes. After a certain number of welds, the resulted weld nugget will drop to below the minimum nugget size required for the weld quality as shown in the graph to the right. The number of welds achievable until the resulted weld nugget sizes dropping to the limit of weld quality is called the "electrode life". This is dependent on the form and material of electrodes, the materials to be welded, surface coatings, and the interactions of dynamic welding process parameters.

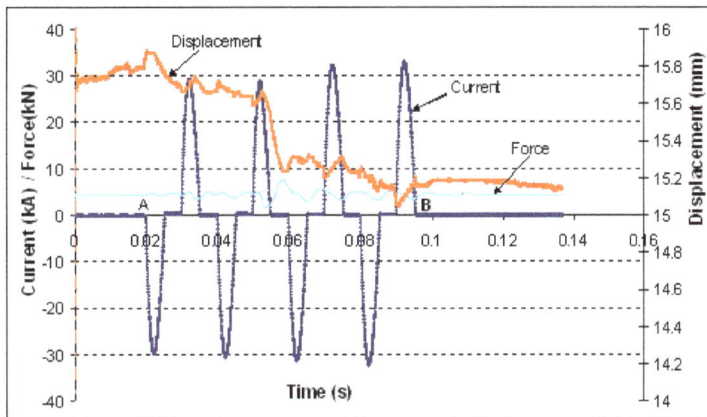

Step Current and Tip-dressing

Apart from adopting new materials and new designs of electrodes, two methods are generally used in production for compensating the electrode degradation in order to maintain the weld quality and increase the electrode life:

- Step current

- Electrode tip-dressing

Step current is a method to plan the spot welding process with a stepwise increasing weld current at each certain number of welds to compensate the loss of current density due to increasing tip diameter as shown in the graph to the right. The higher current needed with the larger tip diameter may be optimized through welding tests or by support of numerical simulations. In this way, more welds can be achieved without replacing electrodes hence a prolonged electrode life.

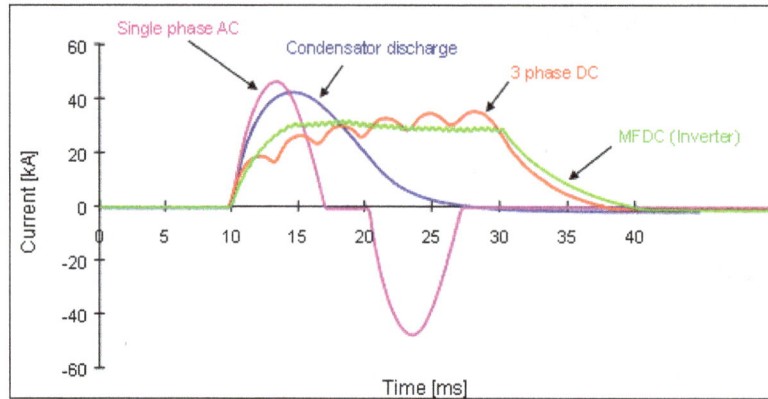

Electrode tip-dressing is a method to mechanically re-shape or abrasively clean the electrode tip after a certain number of welds to preserve nearly the same initial tip diameter and surface conditions. In this way, the welding process can be controlled at the same process parameters or slightly regulated by an adaptive control system to maintain consistent weld quality.

New Developments on Tip-dressing and Optimization

Smart Dress is a new tip-dressing system developed by the European consortium of seven partners including SWANTEC in the EC financed research project on: "Adaptive Tip Dress Control for Automated Resistance Spot Welding".

It will be a fully automated adaptive control system that will optimize, monitor, maintain electrode tip quality with combined mechanical cutting and abrasive cleaning, to maintain weld quality and minimize production line stoppages.

Spot Welding

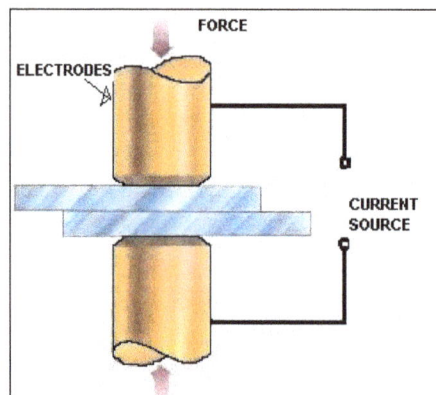

Spot welding is one of the oldest welding processes. It is used in a wide range of industries but notably for the assembly of sheet steel vehicle bodies. This is a type of resistance welding where the spot welds are made at regular intervals on overlapping sheets of metal. Spot welding is primarily used for joining parts that are normally up to 3 mm in thickness. Thickness of the parts to be welded should be equal or the ratio of thickness should be less than 3:1. The strength of the joint depends on the number and size of the welds. Spot-weld diameters range from 3 mm to 12.5 mm.

How Spot Welding Works

Spot welding is one form of resistance welding, which is a method of welding two or more metal sheets together without using any filler material by applying pressure and heat to the area to be welded. The process is used for joining sheet materials and uses shaped copper alloy electrodes to apply pressure and convey the electrical current through the workpieces. In all forms of resistance welding, the parts are locally heated. The material between the electrodes yields and is squeezed together. It then melts, destroying the interface between the parts. The current is switched off and the "nugget" of molten materials solidifies forming the joint.

To create heat, copper electrodes pass an electric current through the workpieces. The heat generated depends on the electrical resistance and thermal conductivity of the metal, and the time that the current is applied. The heat generated is expressed by the equation:

$$E = I^2 * R * t$$

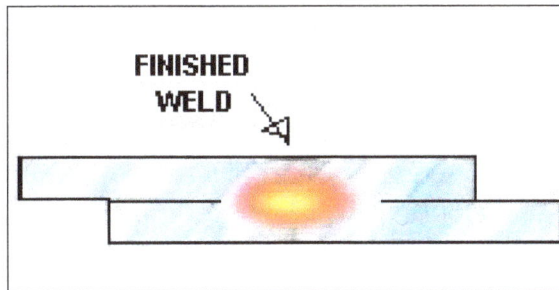

FINISHED WELD

where E is the heat energy, I is the current, R is the electrical resistance and t is the time that the current is applied.

Copper is used for electrodes because it has a low resistance and high thermal conductivity compared to most metals. This ensures that the heat is generated in the workpieces instead of the electrodes.

Materials Suitable for Spot Welding

Steel has a higher electrical resistivity and lower thermal conductivity than the copper electrodes, making welding relatively easy. Low carbon steel is most suitable for spot welding. Higher carbon content or alloy steel tend to form hard welds that are brittle and could crack. Aluminium has an electrical resistivity and thermal conductivity that is closer to that of copper. However, aluminium's melting point is much lower than that of copper, making welding possible. Higher levels of current must be used for welding aluminium because of its low resistivity.

Galvanized steel (i.e. steel coated with zinc to prevent corrosion) requires a different welding approach than uncoated steel. The zinc coating must first be melted off before the steel is joined. Zinc has a low melting point, so a pulse of current before welding will accomplish this. During the weld, the zinc can combine with the steel and lower its resistivity. Therefore, higher levels of current are required to weld galvanized steel.

Flash Welding

Flash welding is a type of resistance welding that does not use any filler metals. The pieces of metal to be welded are set apart at a predetermined distance based on material thickness, material composition, and desired properties of the finished weld. Current is applied to the metal, and the gap between the two pieces creates resistance and produces the arc required to melt the metal. Once the pieces of metal reach the proper temperature, they are pressed together, effectively forge welding them together.

Parameters

According to a study published in Materials and Design, several parameters affect the final product. Flash time is the time that the arc is present. Upset time is the amount of time that the two pieces are pressed together. Flash time needs to be long enough to sufficiently heat the metal before it is pressed together. However, if it is too long, too much of the base metal begins to melt away. The upset time is critical in creating the desired mechanical properties of the finished weld. During the upset, any impurities in the base metal are pressed out creating a perfect weld. If the upset time is too short, all of the impurities may not be pushed out of the base metal creating a defective weld. The upset time is also crucial in the strength of the finished weld because it is during the upset that coalescence occurs between the two pieces of metal. If the upset time is too short, the two pieces of metal may not completely bond.

Very often flash butt welding is controlled by distance rather than time such that the flashing would occur for a pre-determined length, say 5 mm, before the upsetting cycle starts. Upsetting may then also be controlled by distance. A parameter would be set to apply the upsetting force until a certain distance has been upset. It is generally the upsetting distance that is more important than the upsetting time.

At the end of upsetting there is commonly a 'hold time' during which the joint is held still to allow the joint to cool and the two pieces of metal to completely bond.

Applications

According to the Journal of Materials Processing, the railroad industry uses flash welding to join sections of mainline rail together to create continuous welded rail (CWR), which is much smoother than mechanically-joined rail because there are no gaps between the sections of rail. This smoother rail reduces the wear on the rails themselves, effectively reducing the frequency of inspections and maintenance. Continuous welded rail is particularly used on high-speed rail lines because of the smoothness of the rail head. A study published in Materials Science and Design proved that flash welding is also beneficial in the railroad industry because it allows dissimilar metals, including non ferrous metals, to be joined. This allows switches and crossings, which are generally composed of high manganese steel, to be effectively welded to carbon steel rail with the use of a stainless steel insert, while keeping the desired mechanical properties of both the rails and the crossings intact. The ability of this single process to weld many different metals, with simple parameter adjustments, makes it very versatile. Materials and Design also discusses the use of flash welding in the metal building industry to increase the length of the angle iron used to fabricate joists.

The aluminium industry uses flash welding to join aluminium, steel, and copper in various current-carrying conductors called busbars. The steel is used for strength, the copper is used for conductivity, and the aluminium is used for its combination of cost and conductivity.

Resistance Butt Welding

Resistance butt welding is the simplest form of a group of resistance welding processes that involve the joining of two or more metal parts together in a localised area by the application of heat and pressure. The heat is generated within the material being joined by resistance to the passage of a high current through the metal parts, which are held under a pre-set pressure.

The process is used predominantly to make butt joints in wires and rods up to about 16mm diameter, including small diameter chain. The faces of the pieces to be joined may be flat and parallel or profiled in the case of larger sections. This reduces the initial contact area and further concentrates the heating at the interface. The components are clamped in opposing copper dies, with a small amount of stick-out, and abutted under pressure. Current is passed between the dies causing resistance heating of the weld area. The heat generated during welding depends on the current, the duration of the current, and the resistance. As the resistance is highest at the joint interface, heating is most intense in this area. When the material softens, it deforms under the applied load, giving a solid phase forge weld. No melting occurs. The current is terminated once a pre-set upset length has occurred, or the duration of the current is pre-set. The joint is then allowed to cool slightly under pressure, before the clamps are opened to release the welded component. The weld upset may be left in place or removed, by shearing while still hot or by grinding, depending on the requirements.

Present Status

Equipment is well established for joining steel wire and rod up to about 16mm diameter, and narrow strip. Automated dc welding equipment is available for joining wider strip, up to about 300mm wide for automobile road wheel rims, at rates up to about 12 per minute.

Important Issues

Depending on the application area, issues include control of weld quality, production speed, and control of upset shape.

- Weld quality is normally maintained by good process control coupled with consistent end-preparation of materials. Programmed force and current are used for the most demanding applications, such as wheel rims.

- In some applications, such as wire frames, a small, smooth upset is preferred, to avoid the need to remove it. Careful choice of welding parameters allows a balance between adequate strength and minimum upset.

Hardenable steels can be welded and given a subsequent in-machine, local heat treatment to restore weld area toughness. Alloy and stainless steels require more care to ensure that tenacious surface oxides such as chromium oxide are sufficiently dispersed from the interface during welding.

Benefits

Resistance butt welding is a high speed, clean process and is preferred to flash welding for many small components.

Risks

There are some limitations on component size and geometry: very thin or large sections are unsuitable. The main hazards are (i) the risk of crushing fingers or hands and (ii) burns or eye damage from splash metal.

Seam Welding

Seam welding is a welding technique in which the two similar or dissimilar metals are connected by supplying an electric current and in this process a nugget formation takes place. Mostly, these nuggets are formed in the form of butt or the overlapping welding components. Do you know the meaning of this nuggets? Well, the nuggets are nothing but the small pools of the molten metal which are formed at the location with a high electrical resistance. Seam welding is one of the types of Resistance welding.

Working Principle

The working principle of seam welding is similar to the working principle of resistance welding. According to the working principle of the seam welding, the heat required at the time of the welding is produced due to the resistance of the material. In a simple language, heat generation takes place due to the resistance of the material. Have you ever heard about continuous spot welding? The continuous spot welding is nothing but the seam welding. In the seam welding, we use the electrode of roller type. The two rollers have an identical size. Here, these roller type electrodes are brought in contact with the workpiece. Then, the electric current is supplied to these rollers. As the supplied current is very high the interface surface between the roller and the workpiece starts to melt and thus, a strong weld joint is formed. After that, these rollers start to rotate on the surface of the workpiece. As these rollers move, a generation of a continuous joint is formed. Are you curious to know about the welding speed in case of the seam welding? Well, the welding speed is 60 in/min in this welding technique. This speed is assumed to be standard but if you consider the practical applications then, there are chances that this speed may increase or decrease.

Below image will convey you the exact process of the seam welding. Here in the image, you can see that the two roller types electrodes are moving on the plates. And between these two plates, you can see the formation of the weld.

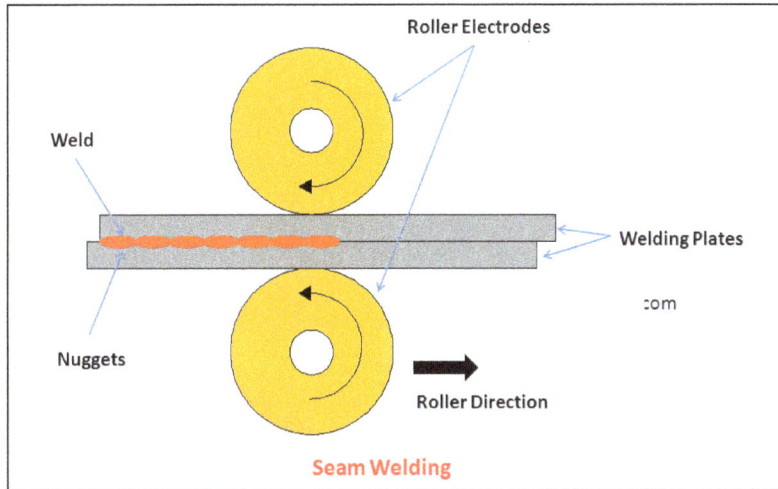

Seam Welding

Precaution

A most important precaution that you must take while performing the seam welding:

Whenever we consider any welding process, then there are many precautions that we have to take in order to make that welding process successful. The same case holds true for the seam welding. In case of a seam welding, you have to take a care about the current that you are supplying. If the supplied current is too high then, the interface between the welding plates and roller will be very hot, in such conditions seam welding cannot be done properly. Also, you must be aware of the welding speed while carrying out this process.

To do the seam welding more effectively, sometimes the weld area is washed with water so that the roller type electrodes will remain cool while entire process.

There are two types of seam welding:

- Intermittent Motion Seam welding.

Here, the roller moves but its speed is not predetermined as that of the continuous motion seam welding.

- Continuous motion seam welding.

In the continuous motion seam welding, the speed of the roller type electrode is always determined first and then, the current is supplied.

Advantages

Let's have a close look at the advantages of this welding process.

- This welding is known for forming the clear weld.

- There are only a few types of the welding processes during which no gas formation takes place or not any other fumes are emitted. And the seam welding is one such type of the welding process.

- Another benefit of this welding process is that there is no requirement of the filler material during this welding method.

- Nowadays, every field in the world is becoming automating. Seam welding can be automated so there is no need of addition labors for doing this process.

- Do you want the simultaneous formation of the single seam weld and parallel seam weld? Then, the seam weld is only for you. This welding process produces the parallel seam weld and the single seam weld at the same time.

- With the help of the continuous spot welding, you can form the gas-tight as well as liquid tight joints.

Disadvantages

- This welding process is very difficult to carry out for the sheets of metals which have a thickness larger than 3 mm. So, it is suggested that use this technique if you have sheets of metals which have a thickness smaller than 3 mm. In simple world, it is not applicable if you have metal pieces above a specific thickness.

- You have to follow a particular path in this welding process. That means that rollers always move in a straight line. So, if you want to make a weld at some complex place then, this process is not applicable.

- There is a need of highly-skilled operator or a machine that can handle the speed of the roller as per the situations.

- The machinery that is used in the seam welding has a very high price. Due to the excessive equipment cost, it is used in a very small proportion.

Applications

- It plays an important role in the manufacturing of all types of the barrels as well as almost every type of the exhaust system.

- Its large number of use is done in case of aircraft tanks, refrigerators, oil transformers, etc.

- It is also used in the welding of the stainless steel, nickel alloys, and magnesium alloys.

Projection Welding

In the projection welding, as per the name, different projections are formed for effective welding. Projection Welding is one of the types of resistance welding and its working principle is quite the same as the resistance welding. The only difference here is that projection or embossed joints are used for the welding purpose.

Working Principle

As per the definition, different projections are formed in this welding technique. Here, the metal pieces that are to be joined are kept in between the two electrodes. A larger pressure force is applied to the electrodes. As current is passed through the system, the heat formation takes place due to the internal resistance of the metal workpieces. One point that you must note down here, is that the heat generation takes place due to the internal resistance of the metal workpieces rather than an electric arc. Those projections concentrate the heat. As the pressure applied to the electrodes increases, this projection collapses and the formation of the fused weld nugget takes place. Thus, a quality weld is formed.

The exact working of the projection welding can be understood by referring the below image.

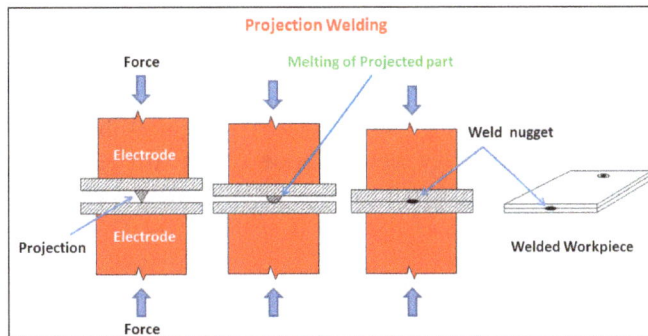

Projection Welding

You can see in the above image that the projection means the embossed joints are formed on one of the base metals and then, these base metals are kept in between the two electrodes and force is applied perpendicular to the electrodes.

But as the applied force increases those sharp projections collapses and the formation of the weld takes place at the weld surface. The above image illustrates the formation of the weld nuggets as well as the collapsing of the sharp projections.

You may have a question that does the shape of the projections is fixed?

No, you can use the projections of any shape. The only precaution that you must take during the selection of the projections is that they must be able to concentrate the heat to form the required weld. The most used types of projection shapes are button-type projections, cone type projections and spherical type projections. Button type projections are used for the metal sheets having higher thickness. This thickness varies in the range of the 24 to 13 gauge while the cone type projections are used for 12 to 5-gauge metal workpieces. And if you have metal sheets of more thickness than the earlier mentioned then, spherical type projection is used.

Characteristics

Following are characteristic of the projection welding that you must know:

- A lot of manufacturers prefer the projection welding instead of other welding techniques. So, do you know the exact reason for it? Well, its answer is quite simple. You can see this welding technique is producing the multiple joints at the same time. So, if you have a very less time and you want the formation of the multiple welds at the same time then, projection welding is a must for you. Therefore, a lot of manufacturers use this welding process instead of others.

- Another important characteristic of the projection welding is that here you get the more electrode life than the spot-welding process. So, it saves your cost of the electrodes and there is no need to change the electrodes in short intervals of time. Some of you might think, why the electrode life is longer in the projection welding than the spot welding? In the projection welding, a weld is formed as a result of the heat and the applied force. But in the case of the spot welding, a weld is formed mostly as a result of the heat generation. So, in the projection welding, electrodes are not affected by the current you are supplying. And mostly due to the force applied the required weld formation takes place. It required very less amount of current. And hence, the life of the electrode in the projection welding is much greater than the spot welding.

- Sometimes, plating material is present on the metal surface and this plating may result in the formation of the cracks while welding. Projection welding removes the plating material efficiently and gives a high-quality weld. Those who want to remove the irritating plating material then, go for this welding technique.

Advantages

Now, it is the time to know about the advantages. So, let's discuss some of the most important benefits of this welding technique:

- As stated above, this welding requires a very small supply of current and thus, it saves the electricity usage. So, less electricity requirement and a longer electrode life are the two most prominent benefits of this welding process.

- While doing spot welding there is a limitation on the thickness of the metal that has to be welded. But in this welding, almost metals of all thickness are welded.

- It can be used effectively for welding of the joints which are on the complicate locations.

- The heat balance is an important part of any welding process and this welding gives a good heat balance while welding.

Disadvantages

- This welding process is not applicable for some types of the coppers and brasses.

- Projection formation is a quite complicated process and it takes time to form the projections. It is very difficult to form the spherical projection and a skilled person is required to form such projections. While making those projections, a height of the projection has to be maintained properly.

- This process is not applicable to all types of workpieces. The composition of the metal workpieces has to be considered while this process and it has some limitations.

Application

As Projection welding is mostly used for the mass production. It has many applications such as:

- Automobile industry uses projection welding to a very large extent.

- This welding process is also used for the fan covers and hollow metal doors.

- It is also used for producing the compressor parts and for the semi-conductors.

- Have you heard about the diamond segment welding? In the diamond segment welding, projection welding finds its applications.

References

- Resistance-welding, technology: swantec.com, Retrieved 16 August, 2019

- Tawfik, David; Mutton, Peter John; Chiu, Wing Kong (2008). "Journal of Materials Processing Technology". Journal of Materials Processing Technology. 196 (1–3): 279–291. Doi:10.1016/j.jmatprotec.2007.05.055

- Spot-welding: robot-welding.com, Retrieved 17 January, 2019

- Zhang, Fucheng; Lv, Bo; Hu, Baitao; Li, Yanguo (2007). "Materials Science and Engineering". Materials Science and Engineering: A. 454-455: 288–292. Doi:10.1016/j.msea.2006.11.018

- What-is-projection-welding-working-principle-advantages-disadvantages-and-application: theweldingmaster.com, Retrieved 18 February, 2019

6
Welding Defects

The weld surface irregularities and imperfections in welded parts are termed as welding defects. These defects can be classified as weld cracking, weld spatter, solid inclusions, porosity etc. The topics elaborated in this chapter will help in gaining a better perspective about these welding defects.

Welding Imperfections

The defects in the weld can be defined as irregularities in the weld metal produced due to incorrect welding parameters or wrong welding procedures or wrong combination of filler metal and parent metal. Weld defect may be in the form of variations from the intended weld bead shape, size and desired quality. Defects may be on the surface or inside the weld metal. Certain defects such as cracks are never tolerated but other defects may be acceptable within permissible limits. Welding defects may result into the failure of components under service condition, leading to serious accidents and causing the loss of property and sometimes also life. Various welding defects can be classified into groups such as cracks, porosity, solid inclusions, lack of fusion and inadequate penetration, imperfect shape and miscellaneous defects.

Porosity

The characteristic features and principal causes of porosity imperfections are described. Best practice guidelines are given so welders can minimise porosity risk during fabrication.

Identification

Porosity is the presence of cavities in the weld metal caused by the freezing in of gas released from the weld pool as it solidifies. The porosity can take several forms:

- Distributed
- Surface breaking pores
- Wormhole
- Crater pipes

Cause and Prevention

Distributed Porosity and Surface Pores

Distributed porosity is normally found as fine pores throughout the weld bead. Surface breaking pores usually indicate a large amount of distributed porosity.

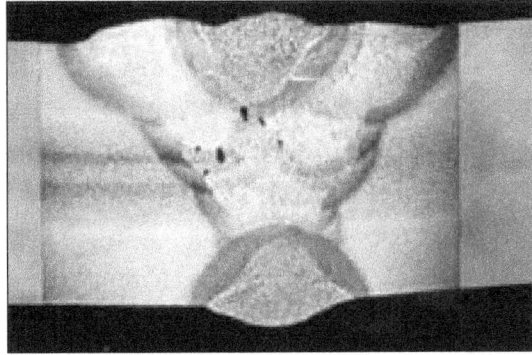

Uniformly distributed porosity.

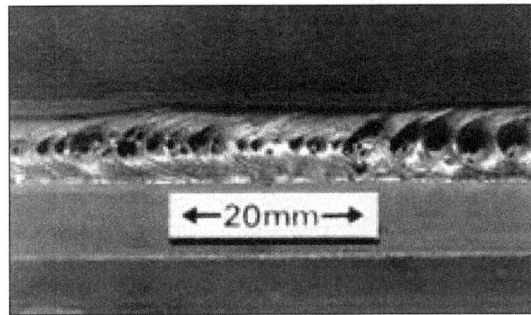

Surface breaking pores (T fillet weld in primed plate).

Cause

Porosity is caused by the absorption of nitrogen, oxygen and hydrogen in the molten weld pool which is then released on solidification to become trapped in the weld metal.

Nitrogen and oxygen absorption in the weld pool usually originates from poor gas shielding. As little as 1% air entrainment in the shielding gas will cause distributed porosity and greater than 1.5% results in gross surface breaking pores. Leaks in the gas line, too high a gas flow rate, draughts and excessive turbulence in the weld pool are frequent causes of porosity.

Hydrogen can originate from a number of sources including moisture from inadequately dried electrodes, fluxes or the workpiece surface. Grease and oil on the surface of the workpiece or filler wire are also common sources of hydrogen.

Surface coatings like primer paints and surface treatments such as zinc coatings, may generate copious amounts of fume during welding. The risk of trapping the evolved gas will be greater in T joints than butt joints especially when fillet welding on both sides. Special mention should be made of the so-called weldable (low zinc) primers. It should not be necessary to remove the primers but if the primer thickness exceeds the manufacturer's recommendation, porosity is likely to result especially when using welding processes other than MMA.

Prevention

The gas source should be identified and removed as follows:

Air Entrainment

- Seal any air leak.
- Avoid weld pool turbulence.
- Use filler with adequate level of deoxidants.
- Reduce excessively high gas flow.
- Avoid draughts.

Hydrogen

- Dry the electrode and flux.
- Clean and degrease the workpiece surface.

Surface Coatings

- Clean the joint edges immediately before welding.
- Check that the weldable primer is below the recommended maximum thickness.

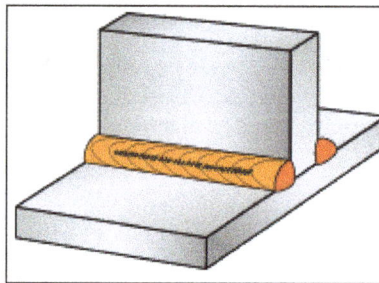

Elongated pores or wormholes.

Wormholes

Characteristically, wormholes are elongated pores which produce a herring bone appearance on the radiograph.

Cause

Wormholes are indicative of a large amount of gas being formed which is then trapped in the solidifying weld metal. Excessive gas will be formed from gross surface contamination or very thick paint or primer coatings. Entrapment is more likely in crevices such as the gap beneath the vertical member of a horizontal-vertical, T joint which is fillet welded on both sides.

When welding T joints in primed plates it is essential that the coating thickness on the edge of the vertical member is not above the manufacturer's recommended maximum, typically 20µm, through over-spraying.

Prevention

Eliminating the gas and cavities prevents wormholes.

Gas Generation

- Clean the workpiece surfaces at and adjacent to the location where the weld will be made.

- Remove any surface contamination, in particular oil, grease, rust and residue from NDT operations.

- Remove any surface coatings from the joint area to expose bright material.

- Check the primer thickness is below the manufacturer's maximum.

Joint Geometry

- Avoid a joint geometry which creates a cavity.

Crater Pipe

A crater pipe forms during the final solidification of the weld pool and is often associated with some gas porosity.

Cause

This imperfection results from shrinkage on weld pool solidification. Consequently, conditions which exaggerate the liquid to solid volume change will promote its formation. Extinquishing the welding arc will result in the rapid solidification of the weld pool.

In TIG welding, autogenous techniques, or stopping the welding wire entering the weld pool before extinquishing the welding arc, will effect crater formation and may promote the pipe imperfection.

Prevention

Crater pipe imperfection can be prevented by controlling the rate at which the welding arc is extinquished or by welder technique manipulating the welding arc and welding wire.

Removal of Stop

- Use run-off tag to enable the welding arc to be extinquisehd outside the welded joint.

- Grind out the weld run stop crater before continuing with the next electrode or depositing the subsequent weld run.

Welder Technique

- Progressively reduce the welding current to reduce the weld pool size (use slope-down or crater fill functions).

- Add filler (TIG) to compensate for the weld pool shrinkage.

Porosity Susceptibility of Materials

Gases likely to cause porosity in the commonly used range of materials are listed in the table.

Principal gases causing porosity and recommended cleaning methods.

Material	Gas	Cleaning
C-Mn steel	Hydrogen, Nitrogen and Oxygen	Grind to remove scale coatings
Stainless steel	Hydrogen	Degrease + wire brush + degrease
Aluminium and alloys	Hydrogen	Chemical clean + wire brush + degrease + scrape
Copper and alloys	Hydrogen, Nitrogen	Degrease + wire brush + degrease
Nickel and alloys	Nitrogen	Degrease + wire brush + degrease

Detection and Remedial Action

If the imperfections are surface breaking, they can be detected using a penetrant or magnetic particle inspection technique. For sub surface imperfections, detection is by radiography or ultrasonic inspection. Radiography is normally more effective in detecting and characterising porosity imperfections. However, detection of small pores is difficult especially in thick sections.

Remedial action normally needs removal by localised gouging or grinding but if the porosity is widespread, the entire weld should be removed. The joint should be re-prepared and re-welded as specified in the agreed welding procedure.

Weld Cracking

One of the primary objectives of any weld fabrication is to prevent weld defects, especially cracks. Cracks are the most severe of all weld defects and are unacceptable in most circumstances. Rework robs the company of precious time and material (that is, money), so prevention is the primary concern.

Cracks don't always happen immediately after welding, and certain cracks, such as the underbead variety, may not be open to the weld surface. Cracks can develop over time after the weld has been subjected to loads while in service. Tensile and fatigue loads; bending, twisting, or flexing; as well as hot and cold expansion and contraction all can occur long after welding, be it two days, two months, or even two years.

The major cause of a crack is when internal stresses exceed the strength of the weld metal, the base metal, or both. And once a focal point for these stresses—that is, a stress riser—develops and accumulates, a crack can propagate.

Defect or Discontinuity

A discontinuity is a weld fault that may or may not be serious enough to cause a rejection. Whether or not it violates code specifications will depend on further examination by a competent person against code requirements or in-house quality assurance specifications. If the fault violates either of these two, it becomes a defect. Defects require repair, but discontinuities do not. Violations of customer requirements often fall under the discontinuity rule and the weld will have to be repaired.

In short, defects always are discontinuities, but not all discontinuities are defects.

Buck Stops

The responsibilities of both welder and supervisor affect weld quality. The welder is responsible for the defect when it is due to his or her skill level or weld deposition technique. Weld characteristics like incomplete fusion, excessively concave or convex bead contours, and improper weld size all can come from poor welding technique, improper travel speed, poor electrode manipulation, incorrect weld parameter settings, as well as failure to notify supervision of a problem with the job at hand.

Supervisors must ensure welders have the tools necessary to do an effective job. They must maintain a shop safety program in compliance with OSHA regulations. They also should, among other things, ensure welders are using the correct base and filler metal; have proper weld procedure testing; work with adequate and functional welding equipment; receive effective and meaningful welder training; and work with properly designed, accessible weld joints.

Responsibility often goes beyond the welder and supervisor, especially when design-for-manufacturability issues come into play. For instance, joint accessibility has become more of a problem nowadays as many designers are not adequately familiar with the requirements of depositing a serviceable, defect-free weld. Can the welder get the gas metal arc welding gun, shielded metal arc welding electrode, or gas tungsten arc welding torch to the work area and still see the joint—or is he welding blind? Does the welder have enough room to manipulate the electrode at all the required angles to deposit a good weld, and still see the joint?

If no design alternative exists, managers must plan for potential weld errors. If an unacceptable weld defect occurs, can a worker get a grinder into the joint to remove the bad weld? If so, how will the weld be repaired? A welder or supervisor can answer all these questions, but the best solution often requires input from customers and product designers.

Crater Cracks

The weld pool has a tremendous amount of built-in stress from weld metal contraction, or shrinkage. Liquid metal is at its maximum expansion, or volume, and so when it cools and solidifies, it has only one direction to go. If the weld pool does not have enough volume after cooling to overcome shrinkage stresses, a crater crack will form, often near the end of a weld, in a high-stress, low-strength area. It's the weld's way of relieving stress.

An excessively concave weld bead contour is a serious candidate for centerline cracking.

The length of the weld deposit also is highly stressed, so that crater crack can very easily travel back through the entire length of the weld centerline. This is a common problem in aluminium and some tool and die steels. The remedy is simple: Fill the crater to its full cross section (the same as the weld size) before the weld is finished. You can accomplish this with various methods. You may pause for two or three seconds at the end of the weld before stopping the arc; or you may choose to backstep (reverse direction of travel) for about 0.5 inch at the end of the bead.

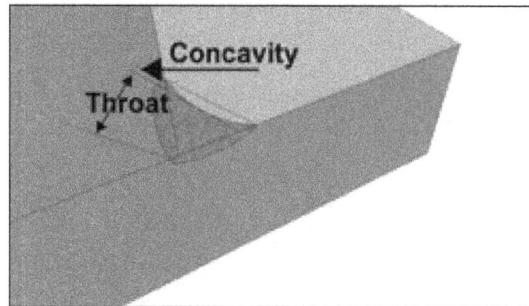

An excessively concave weld bead contour is a serious
candidate for centerline cracking.

Concave Beads

An excessively concave bead profile is a common problem with fillet welds, especially those on stainless steel, INCONEL alloys, and aluminium, but plain carbon steel isn't immune. A certain amount of concavity may be acceptable, depending on the welding requirements. But an excessively concave weld bead contour is a serious candidate for centerline cracking. It generally occurs immediately when welding aluminium (where it's often called "hot cracking"), and is slightly delayed in other materials, after the metal cools to about room temperature.

The problem with concave welds is very similar to that with crater cracks. Reducing the weld throat reduces its strength dramatically, because there isn't enough filler metal in the weld cross section to combat shrinkage stresses. This means those stresses are in control, and a crack develops. If the weld has insufficient throat depth, it probably has insufficient strength.

As with crater cracks, preventing such centerline cracking isn't difficult. Two principal culprits are incorrect travel speed and voltage setting. Voltage is a measure of electrical "pressure," a force pushing down on the face of the liquid weld metal. A small reduction in arc voltage (1 to 1.5 volts) can make a big difference in the weld bead's contour. Reducing the voltage too much, though, can result in a severely convex weld bead contour.

If you set voltage too high, the weld pool becomes difficult to control, and this may encourage you to increase travel speed. This in turn gives you insufficient weld throat depth and weld strength. Once the pool gets ahead of the arc, it's over. You probably will get incomplete penetration, lack of fusion, and undercutting—common problems with vertical-down welding. In fact, performing a vertical-down fillet weld with an acceptable weld throat requires masterful weld pool control. To avoid these problems, slow the travel speed and give the weld time to build an acceptable bead contour.

Convex Beads

Excessively convex bead contours—that is, excessive weld reinforcement—isn't generally associated with weld cracking, though such welds can cause problems. You can waste a lot of time and weld metal depositing an excessively high bead profile. Such a weld is unsightly and almost always unacceptable, mainly because of the weld re-entrant angle to the base metal.

Such bead shapes can have an effect on cracking, especially on cracking that occurs over time. The crack generally is directed down into the base metal, right at the weld toe. If you don't create a smooth transition of weld metal to base metal, you can disrupt the flow of forces through the weld. Such a high volume of weld metal creates significant shrinkage forces. When these forces exceed the strength of the weld, cracking ensues.

To avoid this problem, try increasing travel speed. You can also take a look at your voltage setting. A small increase in voltage increases electrical pressure, forcing the weld contour down to a more acceptable profile.

Undercut Defects

Undercut defects reduce the base metal thickness where the base metal meets the filler metal. This loss of metal interrupts the transfer of stresses from member to member through the weld. If severe, this creates a stress concentration point and has the potential to accumulate and initiate a crack, rapidly.

On high-stress joints, acceptable levels of undercut are near zero. The American Welding Society's D1.1, D1.2, D1.5, and D1.6 codes have extremely low acceptable limits of undercut, depending on the defect's orientation in relationship to the applied stress direction and base metal thickness.

Undercut develops because of improper welding techniques and procedure settings. It usually has no single cause but can come from a range of factors, including incorrect voltage settings, travel speed, and electrode-to-work angle. On fillet welds especially, if voltage (electrical pressure) is too high and the electrode angle favors one member more than another, the arc force will "wash away" the favored member at the weld toe. If the electrode favors one member more and the travel speed is too fast, the arc will naturally melt the member as part of the fusion process, but the high travel speed will not allow the melting electrode to fill in the washed-out area, resulting in an unacceptable weld.

To prevent these defects, make every effort to maintain proper voltage levels. For the constant-voltage processes (nonpulsed GMAW and flux-cored arc welding), the voltage stays fairly constant and can be adjusted manually. For constant-current processes, GTAW, and SMAW, voltage varies with the arc length. If you increase the arc length, you increase arc voltage. Be sure to maintain a correct electrode angle, and try decreasing travel speeds to allow weld deposition to do its job.

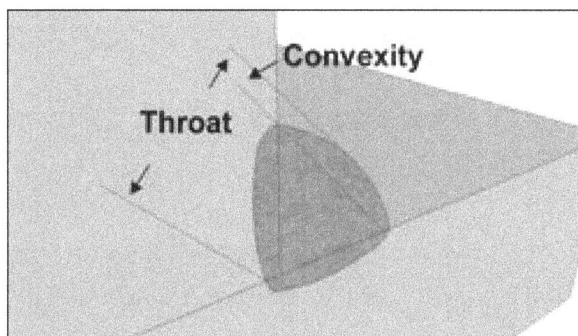

Excessively convex weld bead shapes can have an effect on cracking,
especially on cracking that occurs over time.

Cold Lapping

Overlap or cold lap, is more serious than you might think. If the weld toe remains cold enough that it doesn't fuse with the base metal, the weld just laps over, or lays over the base metal surface without fusing. This leaves no continuity between the weld metal and base metal, so there isn't a path for stress to transfer through the weld into the adjoining member. A classic example of a stress riser, such overlap opens the door for cracking if stress accumulates to unacceptable levels.

Again, the fix isn't difficult. If you don't work the electrode evenly between the two base metals, the weld will favor one member more than the other, and the working parameters (amps and volts) will not liquefy the base metal evenly. Overlap is a common fault when you have to weld blind. Obviously, having to guess where the joint is won't produce favorable results.

The AWS codes call for a "smooth transition" at the toe of the weld. This ensures that the weld stresses flow evenly and, most important, stops those detrimental cracks from forming.

Solid Inclusions

The four common type of solid inclusions usually found in welding are:

- Slag inclusions
- Flux inclusions
- Oxide inclusions
- Metallic inclusions

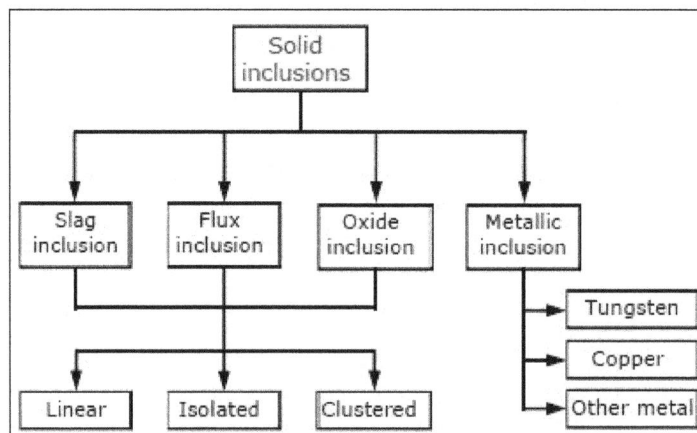

Slag Inclusions

Slag inclusions are slag that is trapped during welding. They have an irregular shape so the appearance is different from a gas pore.

Slag inclusion is caused by an incomplete slag removal from the underlying surface of a multi-pass weld.

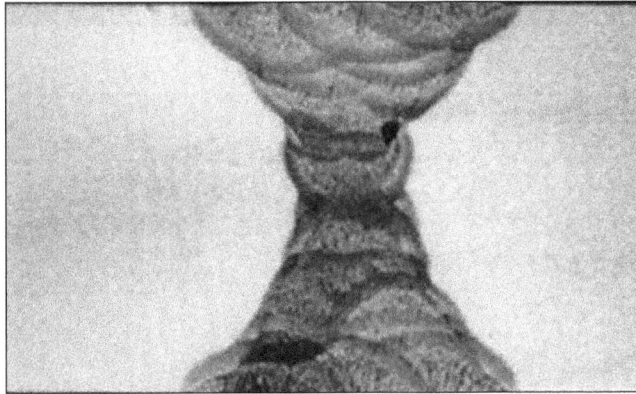

Slag inclusions.

Sometimes the technique used can also cause slag inclusion. For example, when one is welding, the slag flood ahead of the arc.

Some slag can also be entrapment in the work surface. That is why the work surface must be clean and smooth before start welding.

A fine dispersion of inclusions may be present within the weld metal in the MMA process. Although these become a problem only when large or sharp-edged inclusions are produced.

Flux Inclusions

Also differs in appearance from a gas pore, flux inclusions only appear in flux related welding processes (i.e. MMA, SAW, and FCAW).

Flux inclusions are caused by unfused flux due to damaged coating. Sometimes, flux fails to melt and becomes trapped in the weld, particularly in SAW or FCAW process. To overcome this, flux or wire change is recommended.

Furthermore, welding parameters like current and voltage should be adjusted to produce satisfactory welding conditions.

Oxide Inclusions

Cause by oxides trapped during welding, they also have an irregular shape which differs in appearance from a gas pore.

To prevent oxide inclusions, properly grind and cleaned the work surface from heavy mill scale or rust prior to welding.

There is a type of oxide inclusion which is puckering and occurs especially in aluminium alloys. You will find a gross oxide film enfoldment in these alloys due to a combination of unsatisfactory protection from atmospheric contamination and turbulence in the weld pool.

Tungsten Inclusions

During TIG welding, particles of tungsten can become embedded in the weldment.

They appear as a light area on radio-graphs as tungsten is denser than the surrounding metal and absorbs larger amounts of X-ray or gamma radiation.

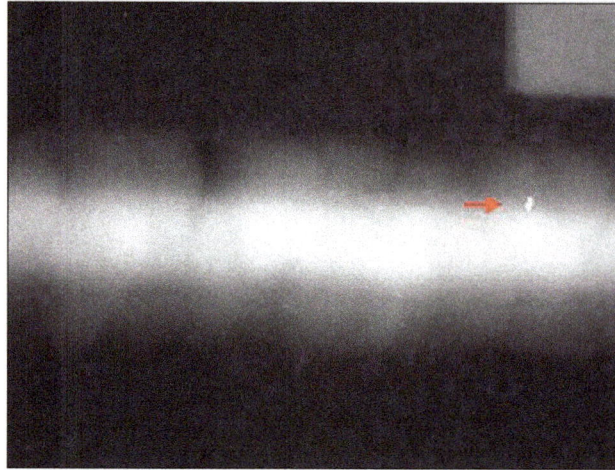

Tungsten inclusions on a radiography film.

Obviously keeping the tungsten out of the weld pool is the main prevention for Tungsten inclusions but sometimes the filler wire touches the tungsten tip which also contaminates the weld pool.

Spatter could also contaminate the tip but this can be avoided by reducing the current and adjusting the shielding gas flow rate.

An overheated electrode due to its extension beyond the normal distance from the collet can also cause tungsten inclusion. If this is the case, reduce the electrode extension and/or the welding current.

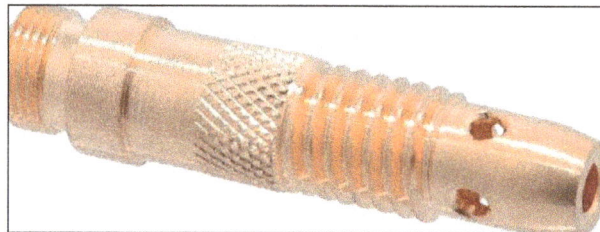

A TIG Collet.

Speaking about collet, before you start welding, make sure the tightening is adequate. So, check your collet and tighten it.

Insufficient shielding gas flow rate or excessive droughts can result in oxidation of the electrode tip. Thus, adjust the shielding gas flow rate, protect the weld area and ensure that the post gas flow after stopping the arc continues for at least five seconds.

Sometimes, Splits or cracks can appear in the electrode. Therefore, to avoid tungsten inclusions, you should change the electrode and ensure the correct tungsten size selected to match the welding current.

Due to the use of argon-oxygen or argon-carbon dioxide mixtures in MAG welding, the shielding gas sometimes becomes insufficient. Therefore, correct the gas composition.

Weld Spatter

Weld spatter consists of droplets of molten metal or non-metallic material that are scattered or splashed during the welding process. These small bits of hot material may fly and fall on the workbench or on the floor, while others may stick to the base material or any surrounding metallic material. They are easy to identify since they are round, small, ball-like substances when they solidify.

Causes of Spatter

Mig welding is characterized by sparks and spatter flying all over the place. It looks great on movies, but when we are doing the welding we realize spatter is a bad thing. It creates more work by increasing clean up time, it wastes material, and it can burn you if you are not wearing the right PPE. It is almost impossible to eliminate spatter in MIG welding, but we can certainly reduce it by understanding what causes it in the first place.

Most of us don't have the option to buy the latest technology in welding equipment to eliminate spatter. We have to weld with what we have. So we are not going to talk about how equipment can help eliminate spatter. This list are items that you can change right now and for free.

- Incorrect settings: procedures that are out of whack will cause spatter. Amperage, voltage and electrical stick out a crucial.

  - Amperage: Amperage in GMAW is determined by your wire feed speed. Running amperage that is too high will cause spatter. To correct either lower the amperage by decreasing the wire feed speed or increase the voltage.

  - Voltage: Per the above, if your voltage is too low your spatter levels will increase. Increase your voltage until spatter decreases.

  - Electric Stick Out (ESO): Electrical stick out is the distance from your contact tip to the work piece. When mig welding you want to be around 3/4″. A bit more for high amperage. Excessive stick out will increase spatter somewhat, but it will create bigger problems (porosity due to lack of shielding gas and lack of penetration).

- Work angle too steep: there is a debate on whether pushing or dragging while mig welding is the way to go. Regardless of which you prefer make sure your drag (pull) or push work angle does not exceed about 15 degrees. At times there is no choice if reach is a problem. But when you can control it do not exceed 15 degrees. Steep angles generate a lot of spatter.

- Surface Contaminants: Rust, oil, paint and other surface contaminants will create spatter. Clean surfaces as best as possible prior to welding.

- Mode of Metal Transfer: Short arc and globular transfers are modes of metal transfer that produce a lot of spatter. To drastically reduce spatter you need to achieve spray transfer. To do this you need a minimum of 83% argon in your shielding mix (a typical mix would be 90/10). However, you also need to be above the transition currents for the diameter of wire you are running. Smaller machines will not be capable of this.

- Erratic Feeding: When the wire feeder cannot feed wire at a constant speed there will be fluctuations in amperage that will drastically affect the arc causing a lot of spatter. Make sure you don't have any feeding issues. For help in correcting this problem take a look at Troubleshooting Erratic Wire Feeding.

- Quality of Consumables: Some applications can live with high levels of spatter, others can't. In robotic applications and other situations in which wire consistency is critical shy away from the cheap-low quality wires. A single spool or drum may be consistent, but across several spools or drums there may be variations in wire diameter, copper coating, and chemistry. Unfortunately AWS allows for such wide range of chemistry that even a coat hanger can be made into a mig wire. The best manufactures keep their own ranges and tolerances and thus produce better product.

- Bad Shielding Gas: This is very uncommon, but shielding gases of low quality can affect spatter levels. What is more common is mislabeling (i.e. getting a 75/25 on a cylinder that has a 90/10 label), but even this is rare. The higher the argon content the smoother the arc. 100% carbon dioxide is cheap and provides good penetration profile, but it creates a lot of spatter.

Common MIG Weld Defects on Aluminium and Steel

Some of the most common weld defects are porosity, lack of fusion and burn through, with aluminium presenting a few more welding challenges than steel. Aluminium conducts heat about six times faster than steel, plus it has excellent thermal conductivity coupled with a low melting point, making it extremely susceptible to warping and burn-through. Additionally, aluminium wire has less tensile strength, which can pose wire feeding issues and lead to weld defects if the correct equipment is not used. We'll highlight the differences here.

Porosity

Shielding gas protects the molten weld pool from the surrounding atmosphere, which would otherwise contaminate the weld. Figure 1 shows how the lack of shielding gas on steel can cause porosity (pinholes) in the weld bead are formed in the face and weld interior in the absence of shielding gas. Lack of shielding gas can be caused by improper setting on the equipment, a hole in the gun liner or wind blowing the shielding gas away.

No Shielding Gas on Steel - A lack of or inadequate shielding gas is easily identified by the porosity and (pinholes) in the face and interior of the weld.

On aluminium, a sooty looking weld can be caused by using a drag vs. a push technique. The soot can be removed, but cutting the weld open will reveal pinholes where impurities are trapped in the weld. Aluminium builds up an oxide that needs to be removed before welding. Wire brushing is the most common method of cleaning aluminium, but it needs to be done with a stainless steel brush to avoid contaminating the weld with the impurities of a steel brush.

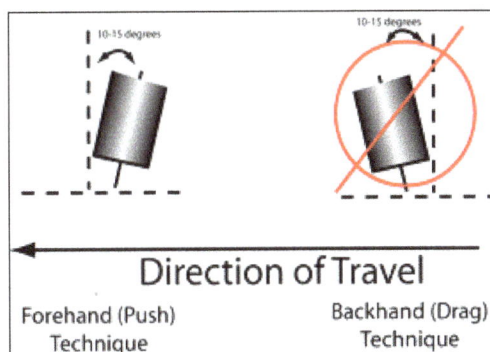

Push vs. drag technique. On steel, either pushing or dragging the gun is acceptable, but with aluminium, the drag technique will lead to weld defects.

Lack of Fusion

Lack of fusion can occur when the voltage or wire feed speed is set too low, or when the operator's travel speed is too fast. Because aluminium conducts heat much faster than steel, it is prone to lack of fusion at the start of a weld until enough energy is put into the weld. Some welding equipment addresses this by automatically ramping up the current at the start of a weld and then decreasing it to avoid too much heat build up.

Craters

With aluminium, craters can form at the end of a weld. If they are not filled in, they create a stress point, which can lead to cracking. This requires the user to quickly trigger the gun again to fill in the crater, although some welding machines offer a crater timer that will fill in the crater when the gun trigger is released.

Burn Through

Too much heat input can be caused by setting voltage or wire feed speed too high or by too slow of a travel speed. This can lead to warping or burn through especially on the thinner materials found in the sign industry, aluminium being more prone to the effects than steel. Generally aluminium requires a faster travel speed than steel to avoid heat build up.

Feeding Aluminium

Because of its low columnar strength, feeding aluminium wire has been likened to pushing a wet noodle through a straw. "Birdnesting," or the tangling of the wire between the drive roll and the liner is a common, time-consuming and costly problem. Clearing it requires the operator to stop welding, cut the wire, discard the wire in the gun, and refeed new wire through the liner. It also may require cleaning or changing the contact tip because of the burnback caused when the wire stops feeding. There are several ways to feed aluminium wire: Push only, spool gun, push-pull system and continuous feed push only system.

- Push only: Feeding aluminium wire through a push only system can be difficult, but it can be done on a limited basis. It requires u-groove drive rolls to provide more surface contact with the wire, a Teflon liner, adequate drive-roll pressure, the ability to keep the gun cable straight and a high tolerance for pain.

- Spool Gun: A spool gun, such as the Spoolmatic 15A or 30A, eliminates the possibility of birdnesting by putting a 4-inch (1-lb.) spool on the gun, so the wire only feeds a few inches. Spool guns can accommodate aluminium wire diameters from .023 to 1/16-inch and allow the operator to use longer cables (15'-50').

- A spool gun needs to have the roll changed after every pound of wire is used, compared with the 8- or 15 lb spool on a push-pull system.

- Push-pull gun: With a push-pull gun, a motor in the gun pulls the wire through the liner, while the motor in the welder or feeder control becomes an assist motor. By maintaining consistent tension on the wire, the push-pull system helps eliminate birdnesting. It is more ergonomic than the spool gun, since the weight of the spool is not in the operator's hands.

Also, the spool needs to be changed less often than on a spool gun and allows the purchase of larger spools. However, remember that aluminium builds up an oxide layer after being exposed to air for a while. If you only go through a pound or two of aluminium a week, the smaller spool may be a better choice.

Choosing the Right Equipment

Choosing the right equipment can address many of these problems before they occur. AutoSetTM technology, as found on many of Miller's Millermatic welders, relieves the operator of having to dial in parameters when welding steel. With Auto-Set, the operator simply dials in the thickness of the steel and the diameter of the wire being used, and the machine sets the optimal voltage and wire feed speed.

For fine tuning, or for welding aluminium, a machine equipped with infinite voltage control, such as the Millermatic 212 AutoSet or Millermatic 252, allows the operator to fine tune settings to avoid putting too much or too little heat into the weld. Older equipment may use tapped settings that can make it more difficult to set, especially when welding thin material.

In addition, both the Millermatic 212 and Millermatic 252 can use spool guns to allow easy set up for both steel and aluminium welding. The Millermatic 252 also features Auto Gun DetectTM, so the operator only has to pick up the MIG or spool gun and pull the trigger to start welding.

For more dedicated aluminium welding, a machine such as the Millermatic 350P, features pulsed MIG welding, which helps eliminate burn through on thin materials. In addition, it features an XR-Aluma-ProTM push-pull gun, which helps eliminate wire feeding issues. Additional technology, such as Aluminium Pulse Hot StartTM provides more power at the start of a weld to avoid the "cold start" to which aluminium welding is prone.

Troubleshooting

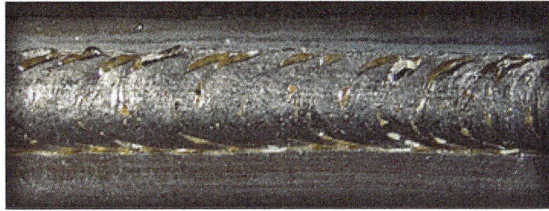

The photo above shows an example of a good weld on steel. Below are photos of a selection of bad welds that can result from a variety of potential problems, including the following:

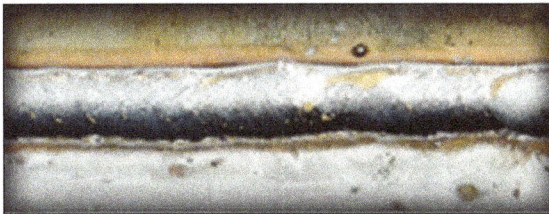

Voltage Too Low - Too little voltage results in poor arc starts, control and penetration. It also causes excessive spatter, a convex bead profile, and poor tie-in at the toes of the weld.

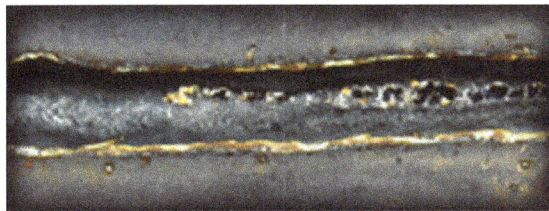

Wire Feed Speed/Amperage Too High: Setting the wire feed speed or amperage too high (depending on what type of machine you're using) can cause poor arc starts and lead to an excessively wide weld bead, burn-through and distortion.

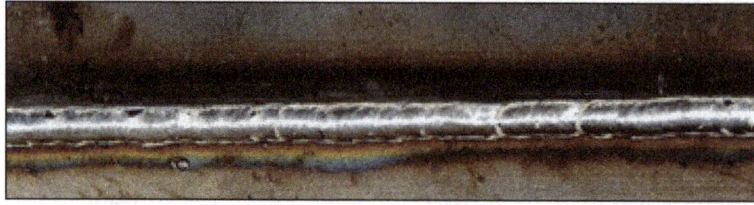

Travel Speed Too Fast: A narrow convex bead with inadequate tie-in at the toes of the weld, insufficient penetration and an inconsistent weld bead are caused by traveling too fast.

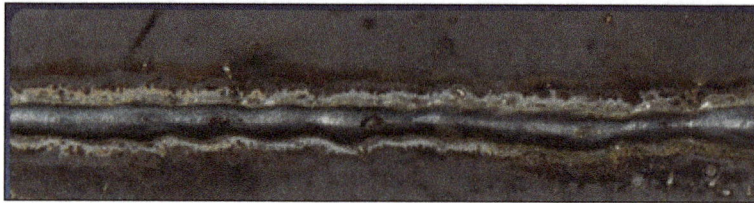

Travel Speed Too Slow: Traveling too slowly may produce a large weld with excessive heat input resulting in heat distortion and possible burn through. In most cases, proper travel speed is when the arc is on the leading edge of the puddle.

Voltage Too High: Too much voltage is marked by poor arc control, inconsistent penetration, and a turbulent weld pool that fails to consistently penetrate the base material.

Tips for Troubleshooting Common MIG Weld Defects

Porosity in Welding

Porosity, one of the most common MIG welding defects, is the result of gas becoming trapped in the weld metal. Inadequate shielding gas coverage is among the biggest culprits, and this can be addressed in several ways. First, check the regulator or flow meter for adequate gas flow, increasing it as necessary. Check the gas hoses and welding gun for possible leaks, and block off the welding area if drafts are present.

To promote proper shielding gas coverage, it's also important to use a large enough nozzle to shield the weld pool fully with gas, keep the nozzle clean and free of spatter, and follow the manufacturer's recommendation for proper contact tip recess.

Other causes of porosity include:

- Dirty base material.

- Excessive gun angle.

- Extending the wire too far from the nozzle. A good rule of thumb is to extend the wire no more than 1/2 inch past the nozzle.

- Wet or contaminated shielding gas cylinders. Replace damaged cylinders immediately.

Lack of Fusion and Cold Lap

The terms cold lap and lack of fusion are often used interchangeably. However, they are slightly different and can happen mil'ndependently or in conjunction with one another in MIG welding.

Lack of fusion is the result of the weld metal failing to fuse completely to the base metal or to the preceding weld bead. It's caused primarily by improper welding gun angle or incorrect travel speed. Avoid this problem by maintaining a 0- to 15-degree gun angle during welding and keeping the arc on the leading edge of the weld pool. It's sometimes necessary to increase travel speed to maintain correct arc position. Insufficient heat in the weld can also cause lack of fusion. This can be remedied by increasing voltage settings or wire feed speeds.

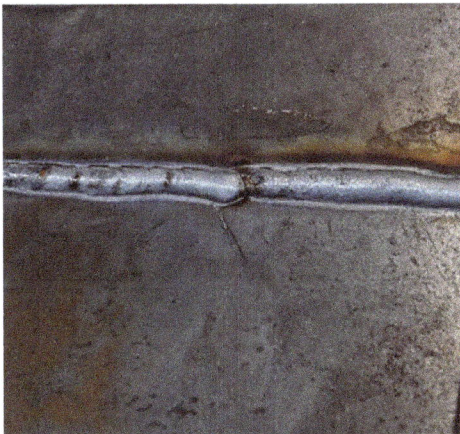

Using incorrect travel speeds can also result in cold lap, which causes the weld to overfill and essentially overlap on the toes of the weld. Increasing travel speed helps prevent this problem.

Burn-through

Burn-through, which happens when the weld metal penetrates completely through the base material, is especially common when welding thin materials less than 1/8 inch or about 12 gauge. Excessive heat is the main cause of burn-through, and this can be fixed by reducing voltage or wire feed speed. Increasing travel speed may also help, particularly when MIG welding on materials especially prone to heat buildup like thin aluminium.

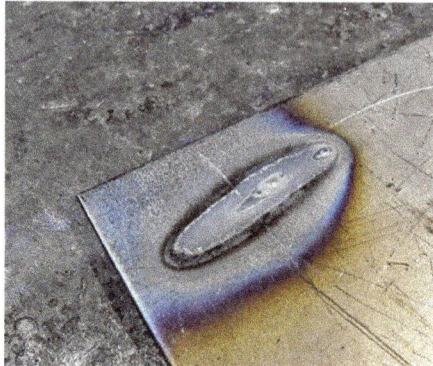

Excessive Spatter

Several issues in the MIG welding process can contribute to excessive spatter, including:

- Insufficient shielding gas.

- Dirty base materials, contaminated or rusty weld wire.

- Voltage or travel speeds that are too high.

- Excessive wire stickout.

Ensuring proper shielding gas flow, cleaning base materials thoroughly, lowering the weld parameter settings and using a shorter stickout are ways to avoid excessive spatter accumulation.

For self-shielded flux-cored wires, be certain to weld with straight polarity (electrode negative) and use a drag technique to minimize the opportunity for spatter buildup. When using flux-cored or metal-cored wires, low voltage can also produce an excessive amount of spatter.

The wrong size contact tip, a worn contact tip or the wrong contact-tip-to-nozzle recess can also cause excessive spatter.

Concave and Convex Weld Beads

The goal is to create a smooth, flat weld bead. Welds that are too concave or convex can compromise the integrity of the finished product.

Concave weld beads are particularly prevalent when welding in vertical-down applications and are simply the result of working against gravity. Adjust the parameters to a lower setting so the weld pool is less fluid and more able to fill in the joint. If a concave weld bead appears in the flat or horizontal position, it's often the result of voltage that's too high, wire feed speed that is too slow or travel speed that is too fast.

Convex weld beads are high, rope-like welds that generally happen in flat and horizontal welding, but can also occur in fillet welds, when the parameters are too cold for the material. Convex weld beads normally have poor fusion of the toes. Increase the voltage to prevent convex beads. Always follow recommended welding procedures and use proper shielding gas for the material, as well as the correct polarity for the wire.

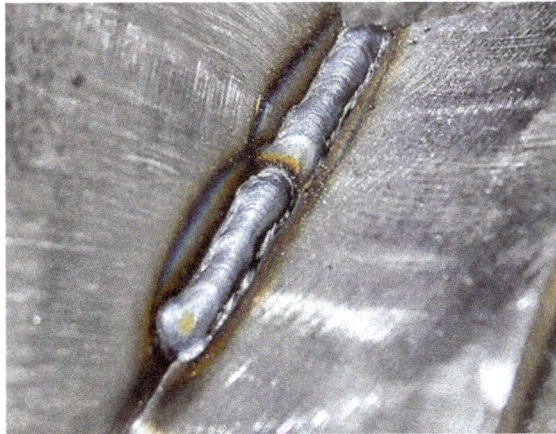

Addressing Common Weld Defects

To minimize the time and money spent addressing MIG weld defects, take a systematic approach for troubleshooting each one should they appear. Look for any variables that have changed during the course of welding — such as parameters or welder technique — then consider these tips as potential remedies.

References

- Defects-imperfections-in-welds-porosity-042, job-knowledge, technical-knowledge: twi-global.com, Retrieved 24 August, 2019

- Cracking-down-on-weld-cracks, arcwelding: thefabricator.com, Retrieved 25 January, 2019

- Solid-inclusions-in-welding: welding-inspectors.com, Retrieved 26 February, 2019

- Causes-of-spatter-and-how-to-eliminate-them: weldinganswers.com, Retrieved 27 March, 2019

- The-most-common-mig-weld-defects-on-aluminum-and-steel-and-how-to-avoid-them: millerwelds.com, Retrieved 28 April, 2019

- Troubleshooting-weld-defects, article-library: millerwelds.com, Retrieved 29 May, 2019

Permissions

All chapters in this book are published with permission under the Creative Commons Attribution Share Alike License or equivalent. Every chapter published in this book has been scrutinized by our experts. Their significance has been extensively debated. The topics covered herein carry significant information for a comprehensive understanding. They may even be implemented as practical applications or may be referred to as a beginning point for further studies.

We would like to thank the editorial team for lending their expertise to make the book truly unique. They have played a crucial role in the development of this book. Without their invaluable contributions this book wouldn't have been possible. They have made vital efforts to compile up to date information on the varied aspects of this subject to make this book a valuable addition to the collection of many professionals and students.

This book was conceptualized with the vision of imparting up-to-date and integrated information in this field. To ensure the same, a matchless editorial board was set up. Every individual on the board went through rigorous rounds of assessment to prove their worth. After which they invested a large part of their time researching and compiling the most relevant data for our readers.

The editorial board has been involved in producing this book since its inception. They have spent rigorous hours researching and exploring the diverse topics which have resulted in the successful publishing of this book. They have passed on their knowledge of decades through this book. To expedite this challenging task, the publisher supported the team at every step. A small team of assistant editors was also appointed to further simplify the editing procedure and attain best results for the readers.

Apart from the editorial board, the designing team has also invested a significant amount of their time in understanding the subject and creating the most relevant covers. They scrutinized every image to scout for the most suitable representation of the subject and create an appropriate cover for the book.

The publishing team has been an ardent support to the editorial, designing and production team. Their endless efforts to recruit the best for this project, has resulted in the accomplishment of this book. They are a veteran in the field of academics and their pool of knowledge is as vast as their experience in printing. Their expertise and guidance has proved useful at every step. Their uncompromising quality standards have made this book an exceptional effort. Their encouragement from time to time has been an inspiration for everyone.

The publisher and the editorial board hope that this book will prove to be a valuable piece of knowledge for students, practitioners and scholars across the globe.

Index